R692

3 1571 00215 5656

S0-CFW-786

SIXTH EDITION

Copyright © 1986, 1991 by Macmillan Publishing Company, a division of Macmillan, Inc.
Copyright © 1983 by The Bobbs-Merrill Co., Inc.
Copyright © 1965, 1970, 1976 by Howard W. Sams & Co., Inc.

All rights reserved. No part of this book may be reproduced or transmitted in any form or by any means, electronic or mechanical, including photocopying, recording, or by any information storage and retrieval system, without permission in writing from the Publisher.

Macmillan Publishing Company   Maxwell Macmillan Canada, Inc.
866 Third Avenue   1200 Eglinton Avenue East, Suite 200
New York, NY 10022   Don Mills, Ontario M3C 3N1

Macmillan Publishing Company is part of the Maxwell Communication Group of Companies.

Production services by the Walsh Group, Yarmouth, ME.

Library of Congress Cataloging-in-Publication Data

Ball, John E.
    Builders math, plans, specifications / by John Ball.—6th ed. / revised
  by John Leeke.
       p.   cm.—(Carpenters and builders library ; v. 2)
  "An Audel book."
  Includes index.
  ISBN 0-02-506452-5
    1. Building—Mathematics.   2. Building—Specifications.
  3. Architectural drawing.   I. Title.   II. Series: Ball, John E.
  Carpenters and builders library ; v. 2.
  TH437.B28   1991
  692—dc20   91–18594
                                                   CIP

While every precaution has been taken in the preparation of this book, the Publisher assumes no responsibility for errors or omissions. Neither is any liability assumed for damages resulting from the use of the information contained herein.

Macmillan books are available at special discounts for bulk purchases for sales promotions, premiums, fund-raising, or educational use. For details, contact:

Special Sales Director
Macmillan Publishing Company
866 Third Avenue
New York, NY 10022

10 9 8 7 6 5 4 3 2 1
Printed in the United States of America

DISCARDED BY THE
GLEN COVE PUBLIC LIBRARY

Carpenters and Builders Library

Volume Two: Builders Math, Plans, Specifications

by John E. Ball
Revised and Edited by John Leeke

"By Hammer and Hand Great Works Do Stand"

An Audel® Book

GLEN COVE PUBLIC LIBRARY
GLEN COVE AVENUE
GLEN COVE, NEW YORK 11542

Macmillan Publishing Company
New York

Maxwell Macmillan Canada
Toronto

Maxwell Macmillan International
New York   Oxford   Singapore   Sydney

Contents

Preface

The Audel CARPENTERS AND BUILDERS LIBRARY is a series of four volumes that cover the fundamental tools, methods, and materials used in carpentry. People familiar with previous editions will note that the material has been thoughtfully and extensively reorganized to fall into a more logical sequence for the reader. This volume continues with more specific techniques and information on materials, mathematics, surveying, and building design.

This series was first published in 1923—a time when most carpentry was done using traditional techniques. Revisions since then dropped most of this early material to make room for modern methods. This revision includes the latest in modern practice as well as drawing from the 1923 edition to bring back the most interesting, valuable, and timeless material from that original work.

The goal of this series is to help you expand your knowledge of carpentry. Knowledge is made up of two parts: information and experience. The Audel CARPENTERS AND BUILDERS LIBRARY gives you a foundation of information. As you work at doing carpentry, you gain the experience that builds your knowledge of the trade.

A new objective of this revision is to provide initial steps in the right direction for carpenters who want to move up in the company they work for or to start their own business. Examples are the use of personal computers for drafting in the Drawing chapter of Volume 1, and the Workshop Design chapter in Volume 4.

As an additional aid to the person who wants to know more about his/her trade, a reference section has been added to each of the volumes. This section gives the reader additional information on resources pertinent to topics in that volume, complete with addresses and phone numbers and a brief description of services provided.

Major revisions in this volume include:

- Expanded hardware chapter includes additional bolts, drives, and screws.
- Expanded chapter on wood combines several previous chapters into a more cohesive unit. Information is also added on milling, defects, selection, decay, as well as other materials, such as plywood, particleboard, hardboard, and oriented strand board and timber.
- Mathematics section expanded to include more examples relevant to carpentry.
- The elements of the design process have been revised and combined into a single unit dealing with specifications, architectural drawings, and building styles. The chapter on reading architectural drawings is new, and gives the builder the basic information on the elements included in architectural drawing and how to understand them.
- Reference section added.

The editor for the current revisions, John Leeke, is an experienced writer as well as a practicing tradesman and contractor. His practical experience ranges from traditional woodworking skills, to modern shop and site practice, to the use of computers in contracting. He also specializes in architectural restoration of historic buildings, and has achieved national stature. His publication history includes contributions to Taunton Press's BUILDERS LIBRARY and CONSTRUCTION TECHNIQUES series; Time-Life's PORCHES, DECKS AND FENCES; IMG Publishing's WALLS AND MOULDINGS. He has recently completed the

PRACTICAL MAINTENANCE MANUAL for the Maine Historic Preservation Commission. In addition to books, Leeke is also a contributing editor for the OLD HOUSE JOURNAL, and he writes a monthly column, "Restoration Primer," for the JOURNAL OF LIGHT CONSTRUCTION. He also publishes and markets his own series of articles on Practical Architectural Preservation.

Leeke's broad range of experience, coupled with that of other professionals, provides a wide range of top-notch resources for the reader.

A note from the revisions editor:

As I grew up learning about the trades in my father's woodworking shop, we often referred to his 1923 edition of Audel's CARPENTERS AND BUILDERS GUIDE. I credit this set of books and my father's guiding hand for giving me a connection with the earlier traditions of carpentry. I have tried to bring that feeling back to this series by drawing from the 1923 edition when the material still provides practical and usable guidance for the modern carpenter.

I would also like to thank the Forest Products Laboratory for permission to excerpt from the WOOD HANDBOOK for the wood chapter, and for the information included in the hardware chapter.

Please feel free to write to me if you have any comment on this current edition. Your ideas and remarks will be used as the next revision is developed.

John Leeke
RR 1–Box 2947
Sanford, Maine 04073

CHAPTER I

Hardware

The strength and stability of any structure depend heavily on the fastenings that hold its parts together. One prime advantage of wood as a structural material is the ease with which wood structural parts can be joined together with a wide variety of fastenings—nails, spikes, screws, bolts, lag screws, drift pins, staples, and metal connectors of various types. For utmost rigidity, strength, and service, each type of fastening requires joint designs adapted to the strength properties of wood along and across the grain, and to dimensional changes that may occur with changes in moisture content.

Nails

Nails are the most common fasteners used in construction.

Up to the end of the Colonial period, all nails used in the United States were handmade. They were forged on an anvil from nail rods, which were sold in bundles. These nail rods were prepared either by rolling iron into small bars of the required thickness or by the much more common practice of cutting plate iron into strips by means of rolling shears.

Just before the Revolutionary War, the making of nails from

these rods was a household industry among New England farmers. The struggle of the Colonies for independence intensified an inventive search for shortcuts to mass production of material entering directly or indirectly into the prosecution of the war; thus came about the innovation of cut nails made by machinery. With its coming, the household industry of nail making rapidly declined. At the close of the 18th century, 23 patents for nailmaking machines had been granted in the United States, and their use had been generally introduced into England, where they were received with enthusiasm.

In France, lightweight nails for carpenter's use were made of wire as early as the days of Napoleon I, but these nails were made by hand with a hammer. The handmade nail was pinched in a vise with a portion projecting. A few blows of a hammer flattened one end into a head. The head was beaten into a countersunk depression in the vise, thus regulating its size and shape. In the United States, wire nails were first made in 1851 or 1852 by William Hersel of New York.

In 1875, Father Goebel, a Catholic priest, arrived from Germany and settled in Covington, Kentucky; there he began the manufacture of wire nails that he had learned in his native land. In 1876, the American Wire and Screw Nail Company was formed under Father Goebel's leadership. As the production and consumption of wire nails increased, the vogue of cut nails, which dominated the market until 1886, declined.

The approved process in the earlier days of the cut-nail industry was as follows: Iron bars, rolled from hematite or magnetic pig, were fagotted, reheated to a white heat, drawn, rolled into sheets of the required width and thickness, and then allowed to cool. The sheet was then cut across its length (its width being usually about a foot) into strips a little wider than the length of the required nail. These plates, heated by being set on their edge on hot coals, were seized in a clamp and fed to the machine, end first. The cut-out pieces, slightly tapering, were squeezed and headed up by the machine before going to the trough.

The manufacture of tacks, frequently combined with that of nails, is a distinct branch of the nail industry, affording much room for specialties. Originally it was also a household industry, and was carried on in New England well into the 18th century. The wire, pointed on a small anvil, was placed in a pedal-operated vise, which clutched it between jaws furnished with a gauge to regulate the length. A certain portion was left projecting; this portion was beaten with a hammer into a flat head.

Antique pieces of furniture are frequently held together with iron nails that are driven in and countersunk, thus holding quite firmly. These old-time nails were made of four-square wrought iron and tapered somewhat like a brad but with a head which, when driven in, held with great firmness.

The raw material of the modern wire nail factory is drawn wire, just as it comes from the wire-drawing block. The stock is low-carbon Bessemer or basic open-hearth steel. The wire, feeding from a loose reel, passes between straightening rolls into the gripping dies, where it is gripped a short distance from its end, and the nailhead is formed by an upsetting blow from a heading tool. As the header withdraws, the gripping dies loosen, and the straightener carriage pushes the wire forward by an amount equal to the length of the nail. The cutting dies advance from the sides of the frame and clip off the nail, at the same time forming its characteristic chisel point. The gripping dies have already seized the wire again, and an ejector flips the nail out of the way just as the header comes forward and heads the next nail. All these motions are induced by cams and eccentrics on the main shaft of the machine, and the speed of production is at a rate of 150 to 500 or more complete cycles per minute. At this stage, the nails are covered with a film of drawing lubricant and oil from the nail machine, and their points are frequently adorned with whiskers—a name applied to the small diamond-shaped pieces stamped out when the point is formed and which are occasionally found on the finished nail by the customer.

These oily nails, in lots of 500 to 5000 pounds, are shaken with sawdust in tumbling barrels from which they emerge bright and clean and free of their whiskers, ready for weighing, packing, and shipping.

The "Penny" System

This method of designating nails originated in England. Two explanations are offered as to how this interesting designation came about. One is that the six penny, four penny, ten penny, etc., nails derived their names from the fact that one hundred nails cost six pence, four pence, etc. The other explanation, which is the more probable of the two, is that one thousand ten-penny nails, for instance, weighed ten pounds. The ancient, as well as the modern, abbreviation for penny is *d*, being the first letter of the Roman coin denarius; the same abbreviation in early history was used for the English pound in weight. The word *penny* has persisted as a term in the nail industry.

Nail Characteristics

Nails are the carpenter's most useful fastener, and a great variety of types and sizes are available to meet the demands of the industry. One manufacturer claims to produce more than 10,000 types and sizes. Some common types are illustrated in Fig. 1–1.

Nails also have a variety of characteristics, that is, different points, shanks, finishes, and material. The following shapes of points are available:

1. Common blunt pyramidal.
2. Long sharp.
3. Chisel-shaped.
4. Blunt, or shooker.

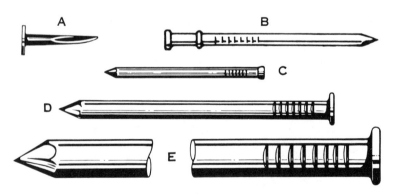

Fig. 1–1. Various nails grouped as to general size. (A) tack; (B) sprig or dowel pin; (C) brad; (D) nail; (E) spike.

Fig. 1-2. Smooth and barbed box nails, lbd size (shown full size). Note the sharp point and thin, flat head.

  5. Side-sloped.
  6. Duck-bill, or clincher.

The heads may be:

  1. Flat.
  2. Oval or oval countersunk.
  3. Round.
  4. Double-headed.

Each of the features or characteristics makes the nail better suited for the job at hand. For example, galvanized nails are weather-resistant, double-headed nails are good for framing where they can be installed temporarily with the second head exposed for easy pulling, barbed nails are good when extra holding power is required.

Tacks—Tacks are small, sharp-pointed nails that usually have tapering sides and a thin, flat head. The regular lengths of tacks range from $\frac{1}{8}$ to $1\frac{1}{8}$ inches. The regular sizes are designated in ounces, according to Table 1-1. Tacks are usually used to secure carpet or fabric.

Table 1-1. Wire Tacks

Size oz.	Length in.	No. per pound	Size oz.	Length in.	No. per pound	Size oz.	Length in.	No. per pound
1	$\frac{1}{8}$	16,000	4	$\frac{7}{16}$	4000	14	$\frac{13}{16}$	1143
$1\frac{1}{2}$	$\frac{3}{16}$	10,666	6	$\frac{9}{16}$	2666	16	$\frac{7}{8}$	1000
2	$\frac{1}{4}$	8000	8	$\frac{5}{8}$	2000	18	$\frac{15}{16}$	888
$2\frac{1}{2}$	$\frac{5}{16}$	6400	10	$\frac{11}{16}$	1600	20	1	800
3	$\frac{3}{8}$	5333	12	$\frac{3}{4}$	1333	22	$1\frac{1}{16}$	727
						24	$1\frac{1}{8}$	666

Fig. 1-3. Brads are small nails. They are used to attach thin strips of wood like moldings. *(Courtesy of The American Plywood Assn.)*

Brads—Brads are small slender nails with small deep heads; sometimes, instead of having a head, they have a projection on one side. There are several varieties adapted to many different requirements. Brad sizes start at about ½ inch and end at 1½ inches; beyond this size they are called finishing nails.

Nails—The term "nails" is popularly applied to all kinds of nails except extreme sizes, such as tacks, brads, and spikes. Broadly speaking, however, it includes all of these. The most generally used are called common nails, and are regularly made in sizes from 1 inch (2*d*) to 6 inch (60*d*). See Table 1–2; Figs. 1–4, 1–5, 1–6.

Spikes—One can think of a spike as an extra large nail, sometimes quite a bit larger. Generally, spikes range from 3 to 12 inches long and are thicker than common nails. Point style varies, but a spike is

Table 1-2. Common Nails

	Plain			Coated		
Size	Length in.	Gauge No.	No. per pound	Length in.	Gauge No.	Per 50-Pound Box
2d	1	15	876	1	16	43,800
3d	$1^{1}/_{4}$	14	568	$1^{1}/_{8}$	$15^{1}/_{2}$	28,400
4d	$1^{1}/_{2}$	$12^{1}/_{2}$	316	$1^{3}/_{8}$	14	15,800
5d	$1^{3}/_{4}$	$12^{1}/_{2}$	271	$1^{5}/_{8}$	$13^{1}/_{2}$	13,500
6d	2	$11^{1}/_{2}$	181	$1^{7}/_{8}$	13	9000
7d	$2^{1}/_{4}$	$11^{1}/_{2}$	161	$2^{1}/_{8}$	$12^{1}/_{2}$	8000
8d	$2^{1}/_{2}$	$10^{1}/_{4}$	106	$2^{3}/_{8}$	$11^{1}/_{2}$	5300
9d	$2^{3}/_{4}$	$10^{1}/_{4}$	96	$2^{5}/_{8}$	$11^{1}/_{2}$	4800
10d	3	9	69	$2^{7}/_{8}$	11	3400
12d	$3^{1}/_{4}$	9	63	$3^{1}/_{8}$	10	3100
16d	$3^{1}/_{2}$	8	49	$3^{1}/_{4}$	9	2400
20d	4	6	31	$3^{3}/_{4}$	7	1500
30d	$4^{1}/_{2}$	5	24	$4^{1}/_{4}$	6	1200
40d	5	4	18	$4^{3}/_{4}$	5	900
50d	$5^{1}/_{2}$	3	14	$5^{1}/_{4}$	4	700
60d	6	2	11	$5^{3}/_{4}$	3	500

normally straight for ordinary uses, such as securing a gutter. However, a spike can also be curved or serrated, or cleft to make extracting or drawing it out very difficult. Spikes in larger sizes are used to secure rails to ties, in the building of docks, and for other large-scale projects.

If you have a very large job to do, it is well to know the holding power of nails. In most instances, this information will not be required, but in more than a few cases it is.

Table 1-3. Withdrawal Force of Cut vs. Wire Nails
(pounds per squre inch)

Wood	Wire Nail	Cut Nail
White Pine	167	405
Yellow Pine	318	662
White Oak	940	1216
Chestnut		683
Laurel	651	1200

Tests for the holding power of nails (and spikes) ranging in size from 6d to 60d are shown in Table 1–4. It is interesting to note, in view of the relatively small force required to withdraw nails, that spikes take tremendous pulling power. In one test it was found that a spike ⅜ inch in diameter driven 3½ inches into seasoned yellow pine required 2000 pounds of force for extraction. And the denser the material, the more difficult the extraction is. The same spike required 4000 pounds of force to be withdrawn from oak and 6000 pounds from well-seasoned locust.

Roofing Nails—The roofing nail has a barbed shank and a large head, which makes it good for holding down shingles and roofing paper felt without damage—the material cannot pull readily through the head.

Such nails come in a variety of sizes but usually ⅜ inch to 1¼ inch long with the nail sized to the material thickness.

Drywall Nails—As the name implies, these are for fastening drywall (Sheetrock™). The shank of the nail is partially barbed and the head countersunk so that if the nail bites into the stud, it takes a good bite. Drywall nails come in a variety of lengths for use with different thicknesses of Sheetrock.

Masonry Nails—Masonry nails are cut, that is stamped, out of a sheet of metal rather than drawn and cut the way wire nails are. A masonry nail is made of very hard steel and is case-hardened. It has a variety of uses but the most common is probably for securing studs or furring to block walls. Safety is important when doing any kind of nailing, but it is particularly important when using masonry nails to wear protective goggles to guard the eyes against flying chips.

Spiral Nails—The most tenacious of all nails in terms of holding power is the spiral nail, also known as the drive screw. Its shank is

Table 1–4. Holding Power of Nails and Spikes (Withdrawal)

Size of spikes Length driven in Pounds resistance to	5 × ¼ in. sq. 4¼ in.	6 × ¼ 5 in.	6 × ½ 5 in.	5 × ⅜ 4¼ in.
drawing, average lbs.	857	857	1691	1202
max. lbs.	1159	923	2129	1556
From 6 to 9 tests each min. lbs.	766	766	1120	687

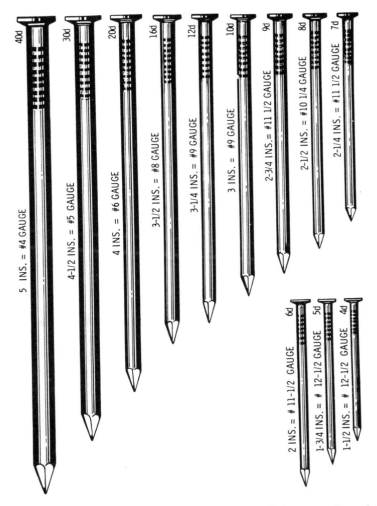

Fig. 1-4. Common wire nails. The standard nail for general use is regularly made in sizes from 1 inch (2*d*) to 6 inches (60*d*).

spiral so that as the nail is driven, it turns and grips the wood. Its main use is to secure flooring, but it is also useful on rough carpentry.

Corrugated Fasteners—This fastener is a small section of corrugated metal with one sharpened and one flat edge. Corrugated fas-

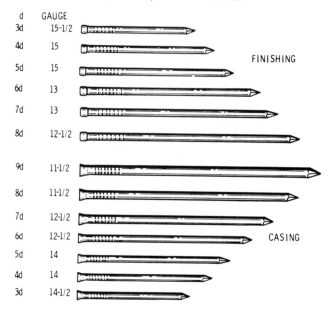

Fig. 1-5. Casing and finishing nails (shown full size). Note the difference in the head shape and size. The finishing nail is larger than a casing nail of equal length, but a casing nail is stronger.

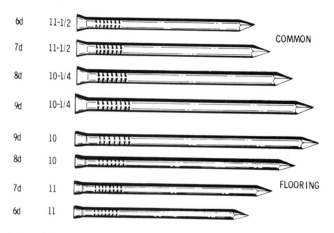

Fig. 1-6. Flooring and common nails (shown full size). Note the variation in head shape and gauge number.

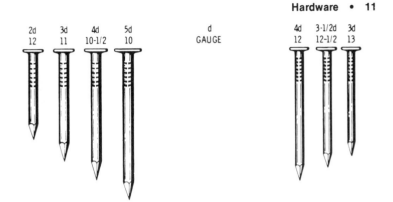

Fig. 1-7. A few sizes of slating and shingle nails. Note the difference in wire gauge.

Fig. 1-8. Hook-head, metal-lath nail. This is a bright, smooth nail with a long, thin flat head, made for application of metal lath. It is also made blued or galvanized.

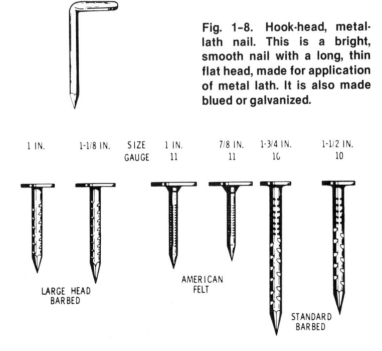

Fig. 1-9. Various roofing nails (shown full size).

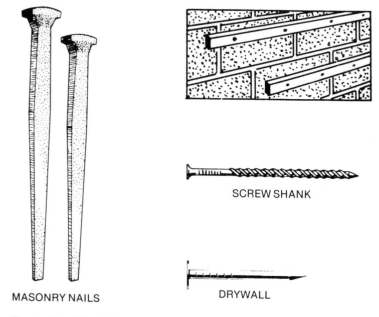

MASONRY NAILS

SCREW SHANK

DRYWALL

Fig. 1-10. Miscellaneous nails.

teners are often used for making boxes or joining wood sections edge to edge. They come in a variety of sizes.

Staples—Many varieties of staples are available, from ones used to secure cable and fencing to posts (such staples are always galvanized) to ones used in the various staple guns. Fence staples range in size from $\frac{7}{8}$ inch to $1\frac{1}{4}$ inches, and some are designed—the so-called slash point—so the legs spread when the staple is driven in place; this makes it grip better.

Selecting Nail Size

In selecting nails for jobs, size is crucial. First consideration is the diameter. Short, thick nails work loose quickly. Long, thin nails are apt to break at the joints of the lumber. The simple rule to follow is to use as long and as thin a nail as will drive easily.

Definite rules have been formulated by which to determine the size of nail to be used in proportion to the thickness of the board that is to be nailed:

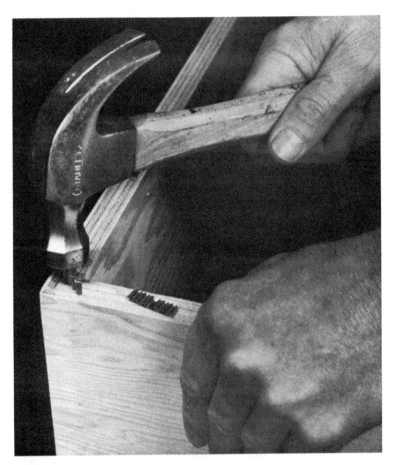

Fig. 1-11. Corrugated nails do not have great holding power, but they are for non-critical or temporary work. *(Courtesy of The American Plywood Assn.)*

1. When using box nails in lumber of medium hardness, the penny of the nail should not be greater than the thickness, in eighths of an inch, of the board into which the nail is being driven.
2. In very soft woods, the nails may be one penny larger, or even in some cases, two pennies larger.

3. In hard woods, nails should be one penny smaller.
4. When nailing boards together, the nail point should penetrate within $\frac{1}{4}$ inch of the far side of the second board.

The kind of wood is, of course, a big factor in determining the size of nail to use: The dry weight of the wood is the best basis for the determination of its grain substance or strength. The greater its dry weight, the greater its power to hold nails. However, the splitting tendency of hard wood tends to offset its additional holding power. Smaller nails can be used in hard lumber than in soft lumber (Fig. 1–12). Positive rules governing the size of nails to be used as related to the density of the wood cannot be laid down. Experience is the best guide.

Driving Nails

In most cases it is not necessary to drill pilot holes for nails to avoid splitting the wood. However, in some instances it is advisable to first drill holes nearly the size of the nail before driving, to guard against it (Fig. 1–13). Also, in fine work, where a large number of nails must be driven, such as applying cedar clapboards, holes should be drilled. This step prevents crushing the wood and possible splitting because of the large number of nails driven through

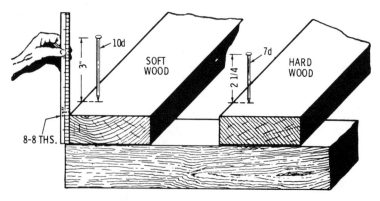

Fig. 1–12. Application of rules 2 and 3 in determining the proper size of nail to use.

Fig. 1-13. To prevent a nail from splitting wood, a pilot hole is sometimes drilled. Pilot-hole drilling is common when using screws. *(Courtesy of The American Plywood Assn.)*

each board. The size of drill for a given size nail should be slightly smaller than the shank diameter.

The right way to drive nails is shown in Fig. 1-14. Fig. 1-15 illustrates the necessity of using a good hammer to drive a nail. The force that drives the nail is due to the inertia of the hammer. This inertia depends on the suddenness with which its motion is brought to rest on striking the nail. With hardened steel, there is practically no give, and all the energy possessed by the hammer is transferred to the nail. On a hammer made with soft and/or inferior metal, all the energy is not transferred to the nail; therefore, the power per blow is less than with hardened steel.

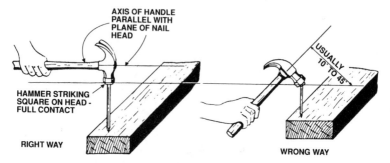

Fig. 1-14. Right way to drive a nail. Hit the nail squarely on the head. The handle should be horizontal when the hammer head hits a vertical nail.

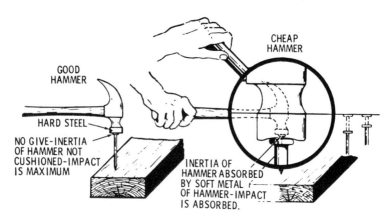

Fig. 1-15. Why a cheap hammer should not be used.

Screws

Wood screws have several advantages over nails. First, screws are harder to pull out. Pull on a screw and pull on a nail—the screw will give greater resistance. Second, should you tire of an item at some future date, screws usually let you disassemble it without great travail. It is possible to damage the work if it is nailed

Table 1–5. Approximate Number of Wire Nails per Pound

American Steel & Wire Co's, Steel Wire Gauge	Length																				
	3/16	1/4	3/8	1/2	5/8	3/4	7/8	1	1 1/8	1 1/4	1 1/2	1 3/4	2	2 1/4	2 1/2	2 3/4	3	3 1/2	4	4 1/2	5
3/8								29	26	23	20	17	15	15	12	11	11	8.9	7.9	7.1	6.4
5/16								43	38	34	29	25	22	20	18	16	15	13	11	10	9.0
1								47	44	40	34	29	26	23	21	20	18	16	14	12	11
2								60	54	48	41	35	31	28	25	23	21	18	16	14	13
3								67	60	55	47	41	36	32	29	27	25	21	18	16	15
4								81	74	66	55	48	41	37	34	31	29	25	22	20	18
5								90	81	74	61	52	45	41	38	35	32	28	24	22	21
6				213	174	149	128	113	101	91	76	65	58	52	47	43	39	34	29	26	24
7				250	205	174	148	132	120	110	92	78	70	61	55	51	47	40	35	31	28
8				272	238	198	174	153	139	126	106	93	82	74	66	61	56	48	42	38	34
9				348	286	238	213	185	170	152	128	112	99	87	79	71	67	58	50	45	41
10				469	373	320	277	242	616	196	165	142	124	111	100	91	84	71	62	55	49
11				510	117	366	323	285	254	233	200	171	149	136	122	111	103	87	77	69	61
12				740	603	511	442	397	351	327	268	229	204	182	161	149	137	118	103	95	87
13			1356	1017	802	688	590	508	458	412	348	297	260	232	209	190	175	153	138	123	110
14		2293	1664	1290	1037	863	765	667	586	536	459	398	350	312	278	256	233	201	176	157	140
15		2899	2213	1619	1316	1132	971	869	787	694	578	501	437	390	351	317	290	256	220	196	177
16		3932	2770	2142	1708	1414	1229	1099	973	872	739	635	553	496	452	410	370	318	277	248	226
17		5316	3890	2700	2306	1904	1581	1409	1253	1139	956	831	746	666	590	532	486	418	360	322	295
18		7520	5072	3824	3130	2608	2248	1976	1760	1590	1338	1150	996	890	820	740	680	585	507	448	412
19		9920	6860	5075	4132	3508	2816	2556	2284	2096	1772	1590	1390	1205	1060	970	895	800			
20	18620	14050	9432	7164	5686	4795	4230	3596	3225	2893	2412	2070	1810	1620	1450	1315	1215	1035			
21	23260	17252	12000	8920	7232	6052	5272	4576	4020	3640	3040	2665	2310	2020	1830						
22	28528	21508	14676	11776	9276	7672															
23	35864	27039	18026	13519	10815	9013															
24	44936	34018	22678	17008	13607	11339															
25	57357	43243	28828	21622	17297	14414															

These approximate numbers are an average only, and the figures given may be varied either way, by changes in the dimensions of the heads or points. Brads and on-head nails will run more to the pound than table shows, and large or thick-headed nails will run less.

together and you want to take it apart. These advantages cost more in the effort and time it takes to install screws.

Screws are normally used to fasten things such as hinges, knobs, etc. to structures, and in the assembly of various wood parts. They are not used in heavy building simply because in this type of work things are built so that there is a minimum of stress on the fasteners and the withdrawal resistance is not required. Indeed, if stress were created, even the most tenacious screw could not stand up much better than a nail—which is to say very little.

The wood screw consists of a gimlet point, a threaded portion, and a shank and head, which may be straight slot or Phillips. More about these later.

Screws of many types are made for specialized purposes, but stock wood screws are usually obtainable in either steel or brass, and, more rarely, are made of high-strength bronze. Three types of heads are standard: the flat countersunk head, with the included angle of the sloping sides standardized at 82°; the round head, whose height is also standardized, but whose contour seems to vary slightly among the products of different manufacturers; and the oval head, which combines the contours of the flat head and the round head. All of these screws are available with the Phillips slot, or crossed slots, as well as the usual single straight slot.

The Phillips slot allows a much greater driving force to be exerted without damaging the head, and it is more sightly than the usual straight-slotted head. By far the greater part of all wood screws used, probably 75% or more, are of the flat-head type.

Material

For ordinary purposes, steel screws, with or without protective coatings, are commonly used. In boat building or other such work where corrosion will probably be a problem if screws are used, the screws should be of the same metal or at least the same *type* of metal as the parts they contact. While it is possible and indeed probable that a single brass screw driven through an aluminum plate, if it is kept dry, will show no signs of corrosion, many brass screws driven through the aluminum plate in the presence of water or dampness will almost certainly show signs, perhaps serious signs, of galvanic corrosion.

Dimensions of Screws

When ordering screws, you must know that length varies with head type. The overall length of a 2-inch flat-head screw is not the same as a 2-inch round-head screw (Fig. 1–16).

Shape of the Head

You can find a variety of head shapes on screws, but the three standard shapes are flat, round, and oval. These usually will more than suffice.

All of these heads are available in either the straight-slotted or Phillips type.

The other forms may be regarded as special or semispecial, that is, carried by large dealers only or obtainable only on special order.

Flat heads are necessary in some cases, such as on door hinges, where any projection would interfere with the proper working of

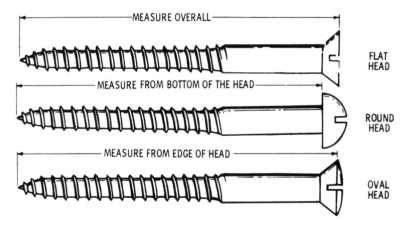

Fig. 1–16. Various wood screws and how their length is measured.

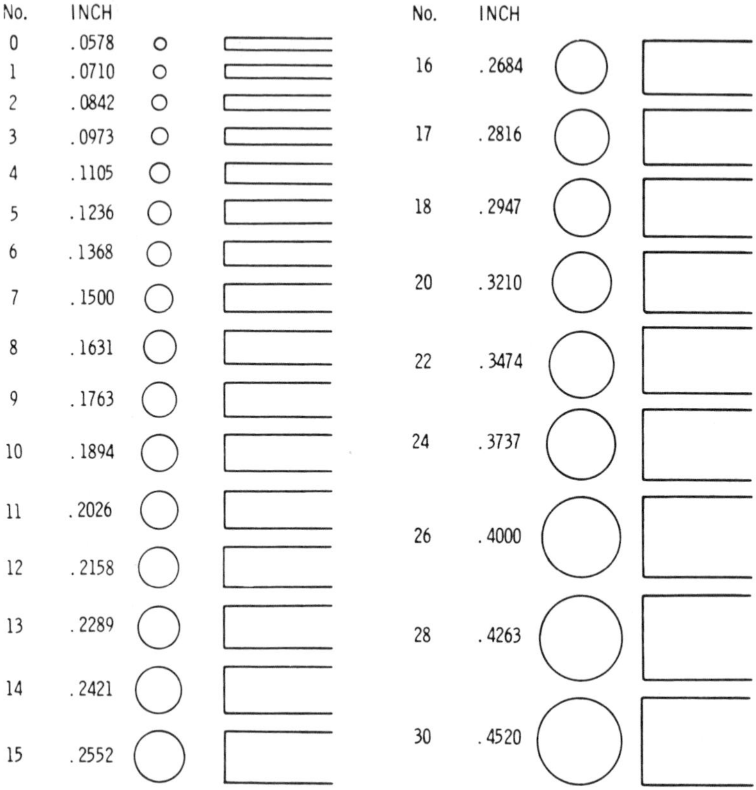

No.	INCH			No.	INCH		
0	.0578						
1	.0710			16	.2684		
2	.0842						
3	.0973			17	.2816		
4	.1105						
5	.1236			18	.2947		
6	.1368						
7	.1500			20	.3210		
8	.1631						
9	.1763			22	.3474		
10	.1894						
11	.2026			24	.3737		
12	.2158			26	.4000		
13	.2289			28	.4263		
14	.2421						
15	.2552			30	.4520		

Fig. 1-17. Wood screw gauge numbers.

the hinge; flat-head screws are also employed on finish work where flush surfaces are desirable. The round and oval heads are normally ornamental, left exposed.

How to Drive a Wood Screw

Driving wood screws is made easier by drilling shank and pilot holes in the wood. Indeed, this may be the only way to do it. Drill a shank-clearance hole. This hole should be the same size as the shank diameter of the screw.

Drill a pilot hole. This hole should be equal in diameter to the

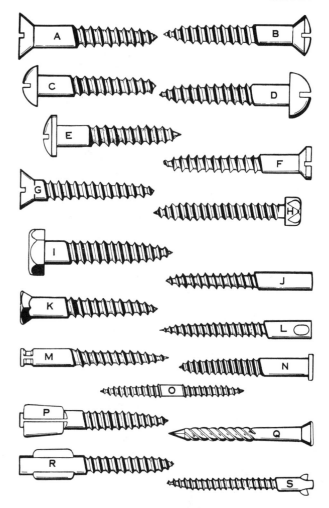

Fig. 1-18. Various wood screws showing the variety of head shapes available. (A) flat head; (B) oval head; (C) round head; (D) piano head; (E) oval fillister head; (F) countersunk fillister head; (G) felloe; (H) close head; (I) hexagon head; (J) headless; (K) square bung head; (L) grooved; (M) pinched head; (N) round bung head; (O) dowel; (P) winged; (Q) drive; (R) winged; (S) winged head. Heads A through G may be obtained with Phillips-type head. Most will never be needed.

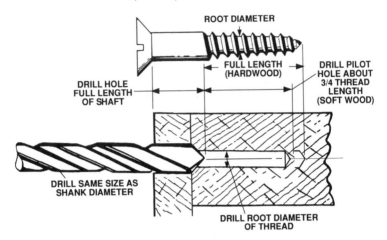

Fig. 1-19. Drilling shank-clearance and pilot holes.

root diameter of the screw thread and about three-quarters of the thread length for soft and medium-hard woods. For extremely hard woods, the pilot-hole depth should equal the thread length.

If the screw being inserted is the flat-head type, the hole should be countersunk (Fig. 1–20).

The foregoing process involves three separate steps. All of these can be performed at once by using a device of the type shown in Fig. 1–21. This tool will drill the pilot hole, the shank-clearance hole, and the countersink all in one operation. Stanley calls its device the Screw-Mate. The Stanley company also makes a Screw-Sink, which counterbores—you can set the head of the screw be-

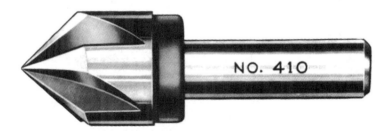

Fig. 1-20. A typical countersink.

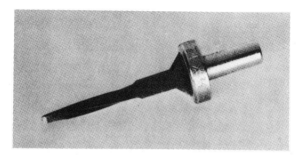

Fig. 1-21. A tool for drilling pilot hole, shank-clearance hole, and countersink in one operation.

Table 1-6. Head Diameters

Screw Gauge	Screw Diameter	Head Diameter		
		Flat	Round	Oval
0	0.060	0.112	0.106	0.112
1	0.073	0.138	0.130	0.138
2	0.086	0.164	0.154	0.164
3	0.099	0.190	0.178	0.190
4	0.112	0.216	0.202	0.216
5	0.125	0.242	0.228	0.242
6	0.138	0.268	0.250	0.268
7	0.151	0.294	0.274	0.294
8	0.164	0.320	0.298	0.320
9	0.177	0.346	0.322	0.346
10	0.190	0.371	0.346	0.371
11	0.203	0.398	0.370	0.398
12	0.216	0.424	0.395	0.424
13	0.229	0.450	0.414	0.450
14	0.242	0.476	0.443	0.476
15	0.255	0.502	0.467	0.502
16	0.268	0.528	0.491	0.528
17	0.282	0.554	0.515	0.554
18	0.394	0.580	0.524	0.580
20	0.321	0.636	0.569	0.636
22	0.347	0.689	0.611	0.689
24	0.374	0.742	0.652	0.742
26	0.400	0.795	0.694	0.795
28	0.426	0.847	0.735	0.847
30	0.453	0.900	0.777	0.900

neath the surface, then plug the hole with a wood plug cut with a plug cutter from matching wood. These tools are made in many sizes, one for each screw size, and they are available in complete sets or separately. The screw size is marked on the tool.

Strength of Wood Screws

Table 1–7 gives the safe resistance, or safe load (against pulling out), in pounds per linear inch of wood screws when inserted across the grain. For screws inserted with the grain, use 60 % of these values.

The lateral load at right angles to the screw is much greater than that of nails. For conservative designing, assume a safe resistance of a No. 20 gauge screw at double that given for nails of the same length, when the full length of the screw thread penetrates the supporting piece of the two connected pieces.

Lag Screws

By definition, a lag screw, shown in Fig. 1–22, is a heavy-duty wood screw provided with a square or hexagonal head so that it may be turned by a wrench. Lag screws are large, heavy screws used where great strength is required, such as for heavy timber work. Table 1–8 gives the dimensions of ordinary lag screws.

How To Put in Lag Screws—First, bore a hole slightly larger than the diameter of the shank to a depth that is equal to the length that the shank will penetrate (Fig. 1–23). Then bore a second hole at

Table 1–7. Safe Loads for Wood Screws

Kind of Wood	Gauge Number							
	4	8	12	16	20	24	28	30
White oak	80	100	130	150	170	180	190	200
Yellow pine	70	90	120	140	150	160	180	190
White pine	50	70	90	100	120	140	150	160

Fig. 1-22. Ordinary lag screw.

the bottom of the first hole equal to the root diameter of the threaded shank and to a depth of approximately one-half the length of the threaded portion. The exact size of this hole and its depth will, of course, depend on the kind of wood; the harder the wood, the larger the hole.

The resistance of a lag screw to turning is enormous when the hole is a little small, but this can be considerably decreased by smearing the threaded portion of the screw with beeswax.

Strength of Lag Screws—Table 1–9 gives the safe resistance, to pull out load, in pounds per linear inch of thread for lag screws when inserted across the grain.

Table 1-8. Lag Screws (inches)

Length	Diameter
3	$5/16$ to $7/8$
$3^1/2$	$5/16$ to 1
4	$5/16$ to 1
$4^1/2$	$5/16$ to 1
5	$5/16$ to 1
$5^1/2$	$5/16$ to 1
6	$5/16$ to 1
$6^1/2$	$7/16$ to 1
7	$7/16$ to 1
$7^1/2$	$7/16$ to 1
8	$7/16$ to 1
9	$7/16$ to 1
10	$1/2$ to 1
11	$1/2$ to 1
12	$1/2$ to 1

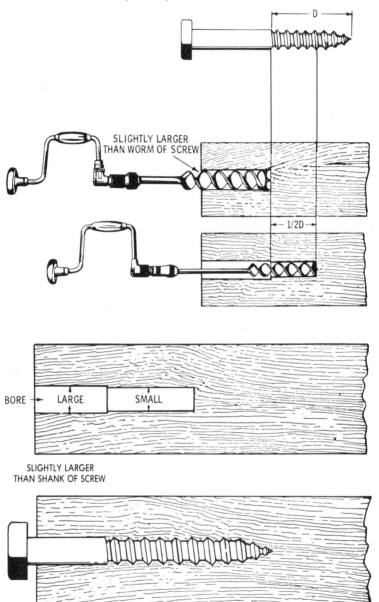

Fig. 1-23. Drilling holes for lag screws.

Table 1-9. Safe Loads for Lag Screws (Inserted across the grain)

Kind of Wood	Diameter of Screw in Inches				
	$1/2$	$5/8$	$3/4$	$7/8$	1
White pine	590	620	730	790	900
Douglas fir	310	330	390	450	570
Yellow pine	310	330	390	450	570

Bolts

Bolts are used to bind parts tightly together where high strength is needed.

Manufacture of Bolts

The bolt-and-nut industry in America was started on a small scale in Marion, Connecticut, in 1818. In that year, Micah Rugg, a country blacksmith, made bolts by the forging process. The first machine used for this purpose was a device known as a heading block, which was operated by a foot treadle and a connecting lever. The connecting lever held the blank while it was being driven down into the impression in the heading block by a hammer. The square iron from which the bolt was made was first rounded so that it could get into the block.

At first, Rugg only made bolts to order, and charged at the rate of 16 cents apiece. This industry developed quite slowly until 1839 when Rugg went into partnership with Martin Barnes. Together they built the first exclusive bolt-and-nut factory in the United States.

Bolts were first manufactured in England in 1838 by Thomas Oliver of Darlston, Staffordshire. His machine was built on a somewhat different plan from that of Rugg's, but no doubt was a further development of the first machine. Oliver's machine was known as the "English Oliver."

The construction of the early machines was carefully kept secret. It is related that in 1842, a Mr. Clark had his bolt-forging ma-

Table 1-10. Standard Wood Screw Proportions

Screw Numbers	A	B	C	D	Number of Threads per Inch
0	....	....	....	0.0578	30
1	....	....	....	0.0710	28
2	0.1631	0.0454	0.030	0.0841	26
3	0.1894	0.0530	0.032	0.0973	24
4	0.2158	0.0605	0.034	0.1105	22
5	0.2421	0.0681	0.036	0.1236	20
6	0.2684	0.0757	0.039	0.1368	18
7	0.2947	0.0832	0.041	0.1500	17
8	0.3210	0.0809	0.043	0.1631	15
9	0.3474	0.0984	0.045	0.1763	14
10	0.3737	0.1059	0.048	0.1894	13
11	0.4000	0.1134	0.050	0.2026	12.5
12	0.4263	0.1210	0.052	0.2158	12
13	0.4427	0.1286	0.055	0.2289	11
14	0.4790	0.1362	0.057	0.2421	10
15	0.5053	0.1437	0.059	0.2552	9.5
16	0.5316	0.1513	0.061	0.2684	0
17	0.5579	0.1589	0.064	0.2815	8.5
18	0.5842	0.1665	0.066	0.2947	8
20	0.6368	0.1816	0.070	0.3210	7.5
22	0.6895	0.1967	0.075	0.3474	7.5
24	0.7421	0.2118	0.079	0.3737	7
26	0.7421	0.1967	0.084	0.4000	6.5
28	0.7948	0.2118	0.088	0.4263	6.5
30	0.8474	0.2270	0.093	0.4546	6

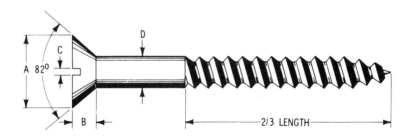

chine located in a room separated from the furnaces by a thick wall. The machine received the heated bars through a small hole cut in the wall; the forge man was not even permitted to enter the room.

Kinds of Bolts

One commonly used bolt is the carriage bolt, which got its name from its prime early use: assembling horsedrawn carriages.

To install a carriage bolt, a hole equal to the diameter of the shank is bored. The bolt is then slipped into the hole, and a hammer is used to pound it down so that the neck seats well in the hole. A nut on the other end completes the job: it can be screwed on without having to hold the other end of the bolt.

Another type of bolt is the machine bolt, used on metal and wood parts. The machine bolt is slipped into the hole and a wrench is used to hold its large square head on one end while another wrench is used to tighten a nut.

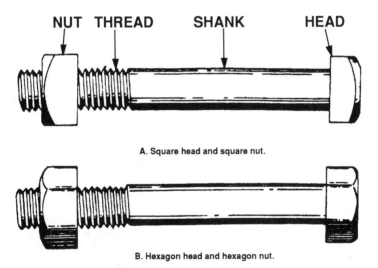

NUT   THREAD   SHANK   HEAD

A. Square head and square nut.

B. Hexagon head and hexagon nut.

Fig. 1–24. Machine bolts.

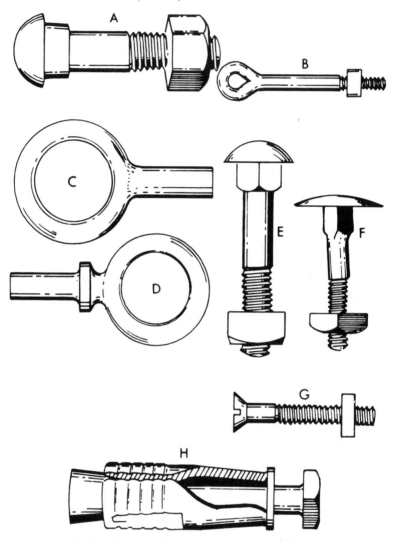

Fig. 1-25.   Various bolts. In the figure, (A) is a railroad track bolt; (B) a welded eye bolt; (C) a plain forged eye bolt; (D) a shouldered eye bolt; (E) a carriage bolt; (F) a step bolt; (G) a stove bolt; (H) an expansion bolt.

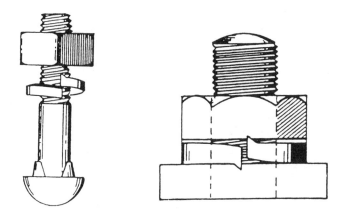

Fig. 1-26. A lock washer. When the nut is screwed onto the bolt, it strikes the rib on the washer, which is much harder than the nut. The rib on the washer is forced into the nut, thus preventing the nut from loosening.

Proportions and Strength of Bolts

Ordinary bolts are manufactured in certain stock sizes. Table 1-11 gives these sizes for bolts from ¼ inch up to 1¼ inches, with the length of thread.

For many years, the coarse-thread bolt was the only type available. Now, bolts with a much finer thread, called the National Fine thread, have become easily available. These have hex heads and hex nuts. They are finished much better than the stock coarse-thread bolts and consequently are more expensive. Cheap rolled-thread bolts, with the threaded portions slightly upset, should not be used by the carpenter. When they are driven into a hole, either the hole is too large for the body of the bolt or the threaded portion reams it out too large for a snug fit. Good bolts have cut threads that have a maximum diameter no larger than the body of the bolt.

When a bolt is to be selected for a specific application, Table 1-13 should be consulted.

Table 1-11. Properties of U.S. Standard Bolts (U.S. Standard or National Coarse Threads)

Diameter	Number of Threads per inch (National Coarse Thread)	Head	Head	Head
$1/4$	20	$3/8$	$13/32$	$1/2$
$5/16$	18	$1/2$	$35/64$	$43/64$
$3/8$	16	$9/16$	$5/8$	$3/4$
$7/16$	14	$5/8$	$11/16$	$53/64$
$1/2$	13	$3/4$	$53/64$	1
$9/16$	12	$7/8$	$31/32$	$1\,15/32$
$5/8$	11	$15/16$	$1\,1/32$	$1\,1/4$
$3/4$	10	$1\,1/8$	$1\,15/64$	$1\,1/2$
$7/8$	9	$1\,5/16$	$1\,29/64$	$1\,47/64$
1	8	$1\,1/2$	$1\,21/32$	$1\,63/64$
$1\,1/8$	7	$1\,11/16$	$1\,55/64$	$2\,15/64$
$1\,1/4$	7	$1\,7/8$	$2\,1/16$	$2\,31/64$
$1\,3/8$	6	$2\,1/16$	$2\,17/64$	$2\,47/64$
$1\,1/2$	6	$2\,1/4$	$2\,31/64$	$2\,63/64$
$1\,5/8$	$5\,1/2$	$2\,7/16$	$2\,11/16$	$3\,15/64$
$1\,3/4$	5	$2\,5/8$	$2\,57/64$	$3\,31/64$
$1\,7/8$	5	$2\,13/16$	$3\,3/32$	$3\,47/64$
2	$4\,1/2$	3	$3\,5/16$	$6\,63/64$

Table 1-12. National Fine Threads

Diameter	Threads per inch
$1/4$	28
$5/16$	24
$3/8$	24
$7/16$	20
$1/2$	20
$9/16$	18
$5/8$	18
$3/4$	16
$7/8$	14
1	14

Table 1-13. Proportions and Strength of U.S. Standard Bolts

Bolt Diameter	Area at Bottom of Threads	Tensile Strength		
		10,000 lbs/in²	12,500 lbs/in²	17,500 lbs/in²
$1/4$	0.027	270	340	470
$5/16$	0.045	450	570	790
$3/8$	0.068	680	850	1190
$7/16$	0.093	930	1170	1630
$1/2$	0.126	1260	1570	2200
$9/16$	0.162	1620	2030	2840
$5/8$	0.202	2020	2520	3530
$3/4$	0.302	3020	3770	5290
$7/8$	0.419	4190	5240	7340
1	0.551	5510	6890	9640
$1 1/8$	0.693	6930	8660	12,130
$1 1/4$	0.890	8890	11,120	15,570
$1 3/8$	1.054	10,540	13,180	18,450
$1 1/2$	1.294	12,940	16,170	22,640
$1 5/8$	1.515	15,150	18,940	26,510
$1 3/4$	1.745	17,450	21,800	30,520
$1 7/8$	2.049	20,490	25,610	35,860
2	2.300	23,000	28,750	40,250

Example—How much of a load may be applied to a 1-inch bolt for a tensile strength of 10,000 pounds per square inch?

Referring to Table 1-13, we find on the line beside 1-inch bolt a value of 5510 pounds corresponding to a stress on the bolt of 10,000 pounds per square inch.

Example—What size bolt is required to support a load of 4000 pounds for a stress of 10,000 pounds per square inch?

$$\text{area at root of thread} = \text{given load} \div 10,000$$
$$= 4000 \div 10,000 = 0.400 \text{ square inch}$$

Referring to Table 1-13, in the column headed "Area at Bottom of Thread," we find 0.419 square inch to be the nearest area; this corresponds to a $7/8$-inch bolt.

Of course, for the several given values of pounds stres, square inch, the result could be found directly from the table, the calculation above illustrates the method that would be er ployed for other stresses per square inch not given in the table.

Example—A butt joint in wood timber with metal fish plates is fastened by six bolts through each member. What size bolts should be used, allowing a shearing stress of 5000 pounds per square inch in the bolts, when the joint is subjected to a tensile load of 20,000 pounds (Fig. 1–27)?

$$\text{load carried per bolt} = 20,000 \div \text{number of bolts}$$
$$= 20,000 \div 6 = 3333 \text{ pounds}$$

Each bolt is in double shear, hence:

$$\text{equivalent single shear load} = \tfrac{1}{2} \text{ of } 3333 = 1667 \text{ pounds}$$

and

$$\text{area per bolt} = \frac{1667}{5000} = 0.333 \text{ square inch}$$

Referring to Table 1–13, the nearest area is 0.302, which corresponds to a ¾-inch bolt. In the case of a dead, or "quiescent," load, ¾-inch bolts would be ample; however, for a live load, take the next larger size, or ⅞-inch bolts.

The example does not give the size of the members, but the as-

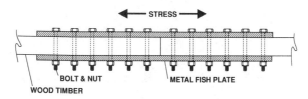

Fig. 1–27.

sumption is they are large enough to carry the load safely. In practice, all parts should be calculated as described in the chapter on the strength of timbers. The ideal joint is one so proportioned that the total shearing stress of the bolts equals the tensile strength of the timbers.

Other Fasteners

There are a variety of specialized fasteners designed to meet particular needs.

Fasteners for Plaster or Drywall

Because of the relatively fragile nature of plaster and drywall in comparison to brick, stone, and concrete, fasteners used with the former must necessarily be different from those used with the latter. Whenever weight of any consequence is involved or a direct outward pull is to be exerted, a fastening is best accomplished with standard wood screws or lag screws inserted through the object to be fastened and driven through the plaster or drywall directly into the studs, rafters, or other framing material beneath. When this is impossible, anchor directly to the plaster or drywall with one or more of the fastening devices discussed in the following paragraphs.

Expansion Anchors—Metal expansion anchors are unsuitable for use with plaster or gypsum board because they tend to crush the walls of the hole into which they are inserted and then fall or pull out easily, assuming they can be tightened in place to begin with. Plastic expansion anchors (Fig. 1–28) are better in this regard, perform their best with radial loads, and are the poorest of any anchor on axial loads. This poor axial-load performance can be countered, to a degree, by using more than one anchor to support the load, as in a ceiling-mounted traverse rod, for example.

Holes for plastic expansion anchors are best bored with a twist or push drill in both plaster and gypsum board in order to get an accurate fit. Holes jabbed with an ice pick, screwdriver, or similar tool are seldom sized correctly for the best friction fit and may have considerable material knocked away from the edge of the hole, inside the wall, making the site useless for an anchor. Bore the hole

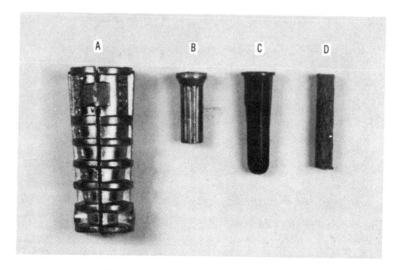

Fig. 1-28. Expansion anchors. (A) made from lead alloy for use with lag screws; (B) made from a softer lead alloy for use with wood screws; (C) made from plastic and best used with sheet-metal screws; (D) made from fiber-jacketed lead—a plug-type anchor sized here for small wood screws.

the diameter specified on the anchor package and use the screw-size specified there, also. The length of the screw should be equal to the length of the anchor plus the thickness of the object to be fastened, as a minimum. As a general rule, sheet-metal screws work better in plastic anchors than do wood screws, possibly because their comparative lack of body taper causes a more effective expansion of the anchor.

Hollow Wall Screw Anchors —These devices are manufactured by a number of different companies and consist of a metal tube having a large flange at one end and an internally threaded collar at the other. A machine screw is inserted through a hole in the flange, extended the length of the tube, and screwed into the threaded collar (Fig. 1-29).

In use, a hole of specified diameter is bored through the gypsum board or plaster and an anchor (Table 1-14) of the proper grip range (depending on the thickness of the drywall or plaster) and

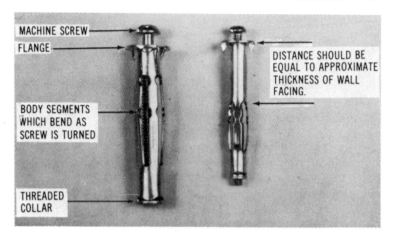

MACHINE SCREW

FLANGE

BODY SEGMENTS WHICH BEND AS SCREW IS TURNED

THREADED COLLAR

DISTANCE SHOULD BE EQUAL TO APPROXIMATE THICKNESS OF WALL FACING.

Fig. 1-29.   Two sizes of hollow wall screw anchors.

Table 1-14.   Allowable Carrying Loads for Anchor Bolts

| Type Fastener | Size | Allowable Load | |
		$^{1}/_{2}''$ Wallboard	$^{5}/_{8}''$ Wallboard
HOLLOW WALL SCREW ANCHORS	$^{1}/_{8}''$ dia. SHORT	50 LBS.	—
	$^{3}/_{16}''$ dia. SHORT	65 LBS.	—
	$^{1}/_{4}''$, $^{5}/_{16}''$, $^{3}/_{8}''$ dia. SHORT	65 LBS.	—
	$^{3}/_{16}''$ dia. LONG	—	90 LBS.
	$^{1}/_{4}''$, $^{5}/_{16}''$, $^{3}/_{8}''$ dia. LONG	—	95 LBS.
COMMON TOGGLE BOLTS	$^{1}/_{8}''$ dia.	50 LBS.	90 LBS.
	$^{3}/_{16}''$ dia.	60 LBS.	120 LBS.
	$^{1}/_{4}''$, $^{5}/_{16}''$, $^{3}/_{8}''$ dia.	80 LBS.	120 LBS.

(Courtesy National Gypsum Co.)

screw size (depending on the weight of the object to be anchored) is inserted so that its length is inside the wall and its flange rests against the wall's surface; the anchor is then lightly tapped with the butt of a screwdriver to seat it and prevent it from turning in the hole. The screw is then turned clockwise with a screwdriver.

As the screw is turned, it draws the collar end of the anchor toward the flange end; four slots cut lengthwise into the tube allow the sections of the tube between the slots to bend outward in response to pressure from the collar until they lie flat against the inside surface of the wall, drawing the flange tightly against the outside surface and locking the anchor securely in place. The screw is then removed, inserted through whatever object is to be fastened, and replaced in the anchor body, an action which can be performed repeatedly without loosening the anchor body.

Hollow wall screw anchors are also manufactured with pointed screws and tapered threaded collars; these can be driven into the wall without drilling a hole first. Very short anchors are also available for use in thin wood paneling and hollow-core flush doors.

Once they are in place, the anchors are removable only with some ingenuity if a large hole in the wall is to be avoided. One method that works, but needs to be done gently, is to replace the screw in the anchor with another one of the same size and thread, but longer. This replacement screw needs to be threaded into the anchor only one turn, if at all. Once it is in place, smack the head of the screw with a hammer. This action may straighten out the bent legs of the anchor so that it can be withdrawn from the wall intact; more frequently, it will either break off the legs or break off the flange. Remove the screw, and in the former case, pull the flanged section of the anchor out of the wall with the fingers (if it won't come out all the way, pull it out as far as possible, cut it in two with diagonal cutters, and let the stubborn half drop down inside the wall cavity). If the flange breaks off, push the remainder of the anchor back inside the wall. The only hole to be patched will be the one originally bored for the anchor, although a too-vigorous hammer blow can produce an additional dimple in the wall surface. The method is a little risky with very soft or very thin wall facings because their relative lack of substance may allow the flange to be driven backward through the facing; let good judgment be the guide—it may be better to leave the anchor alone and learn to live with it.

Toggle Bolts— These differ from hollow wall anchors in that they must be attached to the object to be fastened before they are inserted into the wall. In their larger sizes, they will carry a heavier load (Table 1–14); in all sizes, they are good for axial as well as radial loads, need a larger hole for mounting, and once mounted, cannot be reused if the screw is removed from the toggle. A longer screw is necessary in order to allow the wings of the toggle to unfold in the wall cavity.

Toggle bolts (Fig. 1–30) are simple devices consisting of a center-hinged crosspiece pierced in the middle by a long machine screw. The crosspiece (called the toggle) is composed of two halves (called the wings) hinged around a threaded center through which the screw runs. The wings are normally held at almost a right angle to the screw by spring pressure, but can be folded flat along the screw to allow insertion into the wall. Once inside the wall cavity, they automatically snap upright again (they fold only one way—toward the head of the screw) and prevent removal of the unit. Tightening the screw squeezes the toggle firmly against the inner wall surface and the object to be fastened against the outer wall surface. Removing the screw allows the toggle to drop into the wall cavity; hence, the unit is easily removed, but in most cases, is not reusable because the toggle cannot be recovered.

Toggle bolts are commonly available with screw diameters from ⅛-in. to ⅜-in. and screw lengths to 6 inches, although they

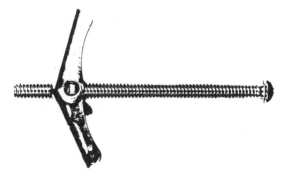

Fig. 1–30. Common toggle bolt.

can be fitted with screws of any maximum practical length by using threaded rod. Minimum screw length should equal the thickness of the object to be fastened, plus the thickness of the wall facing, plus the length of the wings when folded, plus ¼ in. The maximum length should not exceed the minimum by much, if at all, or the screw can bottom against the opposite wall facing and be impossible to tighten.

Hanger Bolts, Dowel Screws, and Toggle Studs—Whenever a hook is to be installed in a plaster or gypsum board ceiling, it should be driven through the ceiling material and into a joist if at all possible. Since most common ornamental hooks are supplied with female machine threads, a device called a hanger bolt (Fig. 1–31) is necessary to accomplish this.

Hanger bolts are relatively short lengths of steel rod having a machine-screw thread on one end and a wood-screw thread on the other. The machine-screw thread is turned into the ornamental hook until it bottoms, and then the whole unit is turned to sink the wood-screw threads into the ceiling and ceiling joist. A pilot hole

Fig. 1–31. Hanger bolt with wood screw threads at top and machine-screw threads at bottom.

should be bored into the joist to make turning easier and to prevent possible breakage of the hook if too much twisting force is applied. Although ornamental hooks are mentioned here as an example, almost any device can be mounted to the ceiling or wall by using a hanger bolt in a similar manner or by using a standard nut on the machine-screw threads protruding from the wall or ceiling.

Dowel screws, which are identical to hanger bolts except that they have wood-screw thread on both ends, can be used to fasten something of wood to a ceiling or wall in the same way.

If an object having female machine threads cannot be fastened to a ceiling joist or wall stud by a hanger bolt, a toggle stud is used instead. All this amounts to is a toggle bolt without a head on the screw, so that the screw can be turned into the female machine threads in the device to be mounted. A toggle stud will not bear as much weight as a hanger bolt, but should be adequate for a small-to-medium-size flower pot and hanger. It is used the same as a toggle bolt.

Framing Fasteners— Framing fasteners are stamped metal pieces (16 or 18 gauge) with predrilled nail (or screw) holes. You set the fastener between the pieces and drive the nails through the holes to lock the members together. The result is a very strong connection. It should be noted that not all building codes accept them, so their use should be checked out beforehand (Fig. 1–32).

Summary

Nails are the carpenter's most useful fastener, and many nail types and sizes are available to meet the demands of the industry. On any kind of construction work, an important consideration is the type and size of nails to use.

An important factor in selecting nails is size. Long, thin nails will break at the joints of the lumber. Short, thick nails will work loose quickly. The kind of wood is a big factor in determining the size of nail to use.

Wood screws are often used in carpentry because of their advantage over nails in strength. They are used in installing various types of building hardware because of their great resistance to pull-

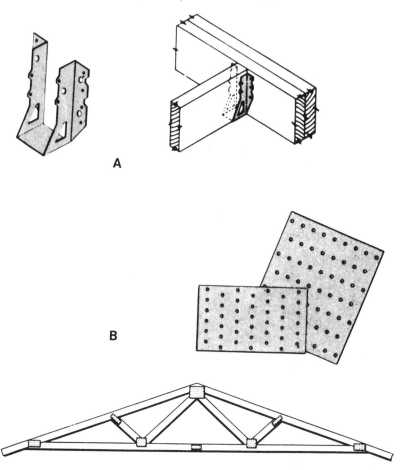

Fig. 1-32. (A) One useful kind of framing fastener is the joist hanger. (B) Perforated plates such as shown can be used to make trusses. *(Courtesy of Teco)*

ing out and because they are more or less readily removed in case of repairs or alterations.

There are generally three standard types of screw heads—the flat countersunk head, the round head, and the oval head, all of which can be obtained in crossed slot, single straight slot, or Phillips slot.

Lag screws or lag bolts are heavy-duty wood screws that are provided with a square or hexagonal head so that they are installed with a wrench. These are large, heavy screws that are used where great strength is needed, such as for heavy timber and beam installations. Holes are generally bored into the wood since the diameter of lag screws is large.

A bolt is generally regarded as a rod having a head at one end and a threaded portion on the other to receive a nut. The nut is usually considered as forming a part of the bolt. Bolts are used to connect two or more pieces of material when a very strong connection is required.

Various forms of bolts are manufactured to meet the demands and requirements of the building trade. The common machine bolt has a square or hexagonal head. The carriage bolt has a round head; the stove bolt has a round or countersunk head with a single slot. Lock washers are used to prevent nuts from loosening. Other fasteners are the toggle, Molly, and expansion bolt.

Review Questions

1. What is nail holding power?
2. Explain the "penny" nail system.
3. What should be considered when selecting a nail for a particular job?
4. Name and describe five kinds of useful nails.
5. Name the three basic head shapes of wood screws.
6. What type of wood screw is used where great strengths is required?
7. What type of head is used on lag screws? Why?
8. What is meant by the root diameter of a screw?
9. What type of head is generally found on a machine bolt?
10. What is meant by "threads per inch"?
11. Explain the purpose of lock washers.
12. What is an expansion bolt?

CHAPTER 2

Wood

Wood is the most versatile, most useful building material, and a general knowledge of the physical characteristics of various woods used in building is important for carpenters and builders.

Growth and Structure

Wood, like all plant material, is made up of cells, or fibers, which when magnified have an appearance similar to, though less regular than, that of the common honeycomb. The walls of the honeycomb correspond to the walls of the fibers, and the cavities in the honeycomb correspond to the hollow or open spaces of the fibers.

Softwoods and Hardwoods

All lumber is divided as a matter of convenience into two great groups: softwoods and hardwoods. The softwoods in general are the coniferous or cone-bearing trees, such as the various pines, spruces, hemlocks, firs, and cedar. The hardwoods are the noncone-bearing trees, such as the maple, oak, and poplar. These terms are used as a matter of custom, for not all so-called softwoods are soft nor are all so-called hardwoods necessarily hard. As a matter of fact, such so-called softwoods as long-leaf southern pine and Douglas fir are much harder than poplar, basswood, etc., which are called hardwoods.

Other and perhaps more accurate terms often used for these two groups are the needle-bearing trees and the broad-leaved trees, referring to the softwoods and hardwoods, respectively. In general, the softwoods are more commonly used for structural purposes such as for joists, studs, girders, and posts, while the hardwoods are more likely to be used for interior finish, flooring, and furniture. The softwoods are also used for interior finish and in many cases for floors, but are not often used for furniture.

A tree consists of:

1. Outer bark—living and growing only at the cambium layer. In most trees, the bark continually sloughs away.
2. In some trees, notably hickories and basswood, there are long tough fibers, called bast fibers, in the inner bark. In other trees, such as the beech, these bast fibers are absent.

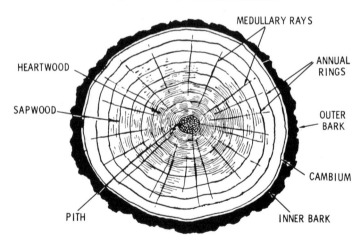

Fig. 2-1. **Cross section of an oak nine years old, showing pith, concentric rings comprising the woody part, the cambium layer, and the bark. The tree grows in concentric rings, or layers, with one layer added each year. The rings are also called annual rings.**

3. Cambium layer. Sometimes this is only one cell thick. Only these cells are living and growing.
4. Medullary rays, or wood fibers, which run radially from the center to the bark.
5. Annual rings, or layers of wood.
6. Pith at the very center.

Around the pith, the wood substance is arranged in approximately concentric rings. The part nearest the pith is usually darker than the parts nearest the bark and is called the heartwood. The cells in the heartwood are dead. Nearer the bark is the sapwood, where the cells carry or store nutrients but are not living.

As winter approaches, all growth ceases, and each annual ring is separate and in most cases distinct. The leaves of deciduous trees, or trees that shed their leaves, and the leaves of some of the conifers, such as cypress and larch, fall, and the sap in the tree may freeze

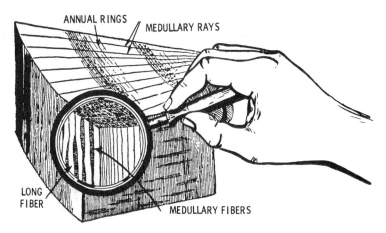

Fig. 2-2. A piece of wood magnified to show its structure. The wood is made up of long, slender cells called fibers, which usually lie parallel to the pith. The length of these cells is often 100 times their diameter. Transversely, bands of other cells, elongated but much shorter, serve to carry sap and nutrients across the trunk radially. Also, in the hardwoods, long vessels or tubes, often several feet long, carry liquids up the tree. There are no sap-carrying vessels in the softwoods, but spaces between the cells may be filled with resins.

hard. The tree is dormant but not dead. With the warm days of the next spring, growth starts again strongly, and the cycle is repeated. The width of the annual rings varies greatly, from 30 to 40 or more per inch in some slow-growing species, to as few as 3 or 4 per inch in some of the quick-growing softwoods.

Lumber Conversion

When logs are taken to the mill, they may be cut in a variety of ways. One way of cutting is quartersawing (Fig. 2–3). Here, each log is ripped into quarters, as shown in the figure. Quartersawing is rarely done this way, though, because only a few wide boards are yielded; there is too much waste. More often, wood is rift-sawed. The log is started as shown in (Fig. 2–4) and is plain sawed until a good figure (pattern) shows, then turned over and rift-sawed. This way there is less waste, and the boards are wide.

The result is *vertical-grain*, or *rift-grain* lumber. Vertical-

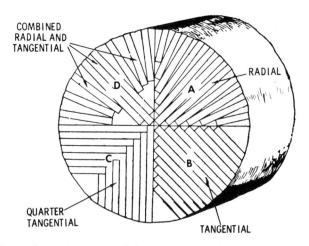

Fig. 2–3. Methods of quartersawing. These are rarely used because waste is extensive.

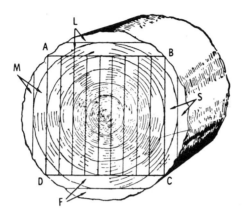

Fig. 2-4. Plain or bastard sawing, sometimes called flat or slash sawing. The log is first squared by removing boards *MS* and *LF,* giving the rectangular section *ABCD.* This is necessary to obtain a flat surface on the log.

grain lumber shrinks less in width and is often used in door stiles and rails because it is less likely to warp.

The plain sawed stock is simply flat-sawn out of the log (Fig. 2-4). This results in *flat-grain* or *side-grain* lumber which is used where shrinkage and warping are less critical.

Seasoning of Wood

Well-developed techniques have been established for removing the large amounts of moisture normally present in green wood. Seasoning is essentially a drying process, but for uses that require them, seasoning includes equalizing and conditioning treatments to improve moisture uniformity and relieve residual stresses and sets. Careful techniques are necessary, especially during the drying phase, to protect the wood from stain and decay and from excessive drying stresses that cause defects and degrade. The established seasoning methods are air drying and kiln drying.

Drying reduces the weight of wood, with a resulting decrease in shipping costs; reduces or eliminates shrinkage, checking, and

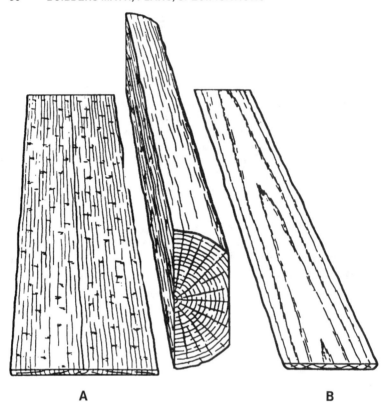

Fig. 2-5. Quartersawed (A) and plainsawed (B) boards cut from a log. *(Courtesy Forest Product Lab)*

warping in service; increases strength and nailholding power; decreases susceptibility to infection by blue stain and other fungi; reduces chance of attack by insects; and improves the capacity of wood to take preservative and fire-retardant treatment and to hold paint.

It is common practice at most softwood sawmills to kiln dry all upper grade lumber intended for finish, flooring, and cut stock. Lower grade boards are often air dried. Dimension lumber is air dried or kiln dried, although some mills ship certain species without seasoning. Timbers are generally not held long enough to be

considered seasoned, but some drying may take place between sawing and shipment or while they are held at a wholesale or distribut-
ing yard. Sawmills cutting hardwoods commonly classify the lumber for size and grade at the time of sawing. Some mills send all
freshly sawed stock to the air-drying yard or an accelerated air-
drying operation. Others kiln dry directly from the green condition. Air-dried stock is kiln dried at the sawmill, at a custom drying
operation during transit, or at the remanufacturing plant before
being made up into such finished products as furniture, cabinet
work, interior finish, and flooring.

Air drying is not a complete drying process, except as preparation for uses for which the recommended moisture content is not
more than 5 percent below that of the air-dry stock. Even when
air-drying conditions are mild, air-dry stock used without kiln drying may have some residual stress and set that can cause distortions
after nonuniform surfacing or machining. On the other hand,
rapid air drying accomplished by low relative humidities produces
a large among of set that will assist in reducing warp during final
kiln drying. Rapid surface drying also greatly decreases the inci-
dence of chemical and sticker stain, blue stain, and decay.

Air drying is an economical seasoning method when carried
out (1) in a well-designed yard or shed, (2) with proper piling
practices (Fig. 2–6), and (3) in favorable drying weather. In cold or
humid weather, air drying is slow and cannot readily reduce wood
moisture to levels suitable for rapid kiln drying or for use.

In kiln drying, higher temperatures and fast air circulation are
used to increase the drying rate considerably. Average moisture
content can be reduced to any desired value. Specific schedules are
used to control the temperature and humidity in accordance with
the moisture and stress situation within the wood, thus minimizing
shrinkage-caused defects. For some purposes, equalizing and conditioning treatments are used to improve moisture content uniformity and relieve stresses and set at the end of drying, so the material
will not warp when resawed or machined to smaller sizes or irregular shapes. Further advantages of kiln drying are the setting of
pitch in resinous woods, the killing of staining or decay fungi or
insects in the wood, and reductions in weight greater than those
achieved by air-drying. At the end of kiln-drying, moisture-
monitoring equipment is sometimes used to sort out moist stock for

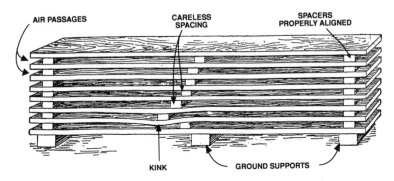

Fig. 2–6. Horizontal stack of lumber for air drying. One end of the pile should be a little higher than the other so that rain water which falls on top or drives into the pile will drain. Each layer should be separated by three or four spacers so that the air will have free access to both sides of each board. The lowest layer should be well elevated above the ground to protect it from dampness. It takes from one to three years to thoroughly season lumber depending upon the character of the wood, climatic conditions, etc. When the spacers are carelessly placed so that they do not lie over each other, the weight must be supported by the board, which especially in the case of a high pile, is considerable, and will in time cause the board to sag resulting in a permanent kink.

redrying and to insure that the material ready for shipment meets moisture content specifications.

Temperatures of ordinary kiln drying generally are between 110° and 180° F. Elevated-temperature (180° to 212° F.) and high-temperature (above 212° F.) kilns are becoming increasingly common, although some strength loss is possible with higher temperatures.

Moisture Content

While the tree is living, both the cells and cell walls are filled with water to an extent. As soon as the tree is cut, the water within the cells, or *free water* as it is called, begins to evaporate. This process continues until practically all of the free water has left the wood. When this stage is reached, the wood is said to be at the fiber-saturation point; that is, what water is contained is mainly in the cell or fiber walls.

Except in a few species, there is no change in size during this preliminary drying process, and therefore no shrinkage during the evaporation of the free water. Shrinkage begins only when water begins to leave the cell walls themselves. What causes shrinkage and other changes in wood is not fully understood; but it is thought that as water leaves the cell walls, they contract, becoming harder and denser, thereby causing a general reduction in size of the piece of wood. If the specimen is placed in an oven which is maintained at 212° F, the temperature of boiling water, the water will evaporate and the specimen will continue to lose weight for a time. Finally a point is reached at which the weight remains substantially constant. This is another way of saying that all of the water in the cells and cell walls has been driven off. The piece is then said to be *oven dry*.

If it is now taken out of the oven and allowed to remain in the open air, it will gradually take on weight, due to the absorption of moisture from the air. As when placed in the oven, a point is reached at which the weight of the wood in contact with the air remains more or less constant. Careful tests, however, show that it does not remain exactly constant—it will take on and give off water as the moisture in the atmosphere increases or decreases. Thus, a piece of wood will contain more water during the humid, moist summer months than in the colder, drier winter months. When the piece is in this condition it is in equilibrium with the air and is said to be *air dry*.

The amount of water contained by wood in the green condition varies greatly, not only with the species but in the same species, and even in the same tree, according to the position in the tree. But as a general average, at the fiber-saturation point, most woods contain from 23 to 30 percent water as compared with the oven-dry weight of the wood. When air dry, most woods contain from 12 to 15 percent moisture.

As the wood dries from the green state, which is that of the freshly cut tree, to the fiber-saturation point, except in a few species, there is no change other than that of weight. It has already

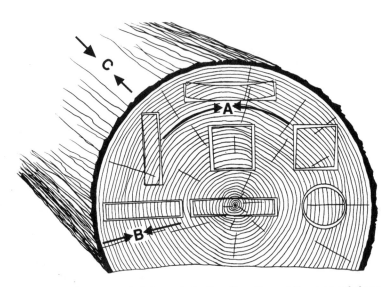

Fig. 2-7. Greatest shrinkage is in the direction of the annual rings (A). Characteristic shrinkage and distortion of flat boards, squares, and rounds as affected by the direction of the annual rings. Tangential shrinkage (A) is about twice as great as radial (B). There is little or no longitudinal (C) shrinkage. *(Courtesy Forest Product Lab)*

been pointed out that as the moisture dries out of the cell walls, in addition to the decrease in weight, shrinkage results in a definite decrease in size (Fig. 2-7). It has been found, however, that there is little or no decrease along the length of the grain, and that the decrease is at right angles to the grain.

This is an important consideration to be remembered when framing a building. For example, a stud in a wall will not shrink appreciably in length, whereas it will shrink somewhat in both the 2-inch and the 4-inch way. In like manner, a joist, if it is green when put in place, will change in depth as it seasons in the building. These principles of shrinkage also explain why an edge-grain or quarter-sawed floor is less likely to open up than a flat-grain floor.

Density

The tree undergoes a considerable impetus early every spring and grows very rapidly for a short time. Large amounts of water are carried through the cells to the rapidly growing branches and leaves at the top of the tree. This water passes upward mainly in the outer layers of the tree. The result is that the cells next to the bark, which are formed during the period of rapid growth, have thin walls and large passages. Later on, during the summer, the rate of growth slows up and the demand for water is less. The cells which are formed during the summer have much thicker walls and much smaller pores. Thus, a year's growth forms two types of wood—the spring wood, as it is called, being characterized by softness and openness of grain, and the summer wood by hardness and closeness, or density, of grain. The spring wood and summer wood growth for one year is called an *annual ring*.

There is one ring for each year of growth. This development of spring wood and summer wood is a marked characteristic of practically all woods that grow in a temperate climate. It is clearly evident in such trees as the yellow pines and firs, and less so in the white pines, maple, and the like. Careful examination will reveal this annual ring, however, in practically all species. It follows, therefore, that a tree in which the dense summer wood predominates is stronger than one in which the soft spring wood predominates. This is a point which should be borne in mind in selecting material for important members such as girders and posts carrying heavy loads. The strength of wood of the same species varies markedly with the density. For example, Douglas fir or southern pine, carefully selected for density, is one-sixth stronger than lumber of the same species and knot limitations in which the spring wood predominates. Trees having approximately one-third or more of cross-sectional area in summer wood fulfill one of the requirements for structural timbers.

Estimating Density—It must be remembered that the small cells or fibers which make up the wood structure are hollow. Wood substance itself has a specific gravity of about 1.5, and therefore will sink in water. It is stated that wood substance of all species is practically of the same density. Strength of wood depends upon its density and varies with its density. The actual dry weight of lumber is a good criterion of its strength, although weight can not always be relied upon as a basis for determining strength, as other important factors frequently must be considered in a specific piece of wood.

The hardness of wood is also another factor which assists in estimating the strength of wood. A test sometimes used is cutting across the grain. This test cannot be utilized in the commercial grading of lumber because moisture content will affect the hardness and because hardness thus measured cannot be adequately defined. The annual rings found in practically all species are an important consideration in estimating density, although the annual rings indicate different conditions in different species. In ring-porous hardwoods and in the conifers, where the contrast between spring wood and summer wood is definite, the proportion of hard summer wood is an indication of the strength of the individual piece of wood. The amount of summer wood, however, cannot always be relied upon as an indication of strength, because summer wood itself varies in density. When cut across the grain with a knife, the density of summer wood may be estimated on the basis of hardness, color, and luster.

In conifers, annual rings of average width indicate denser material or a larger proportion of summer wood than in wood with either wide or narrow rings. In some old conifers of virgin growth, in which the more recent annual rings are narrow, the wood is less dense than where there has been normal growth. On the other hand, in young trees where the growth has not been impeded by other trees, the rings are wider and in consequence the wood less dense. These facts may account for the belief that all second-growth timber and all sapwood are weak. In analyzing wood for density, the contrast between summer wood and spring wood should be pronounced.

Oak, ash, hickory, and other ring-porous hardwoods in general rank high in strength when the annual rings are wide. In this respect they contrast with conifers. These species have more summer wood than spring wood as the rings become wider. For this reason, oak, hickory, ash, and elm of second growth are considered superior because of fast growth and increase in proportion of summer wood. These conditions do not always exist, however; for exceptions occur, especially in ash and oak, where, although the summer wood is about normal, it may not be dense or strong. Very narrow rings in ring-porous hardwoods are likely to indicate weak and brashy material composed largely of spring wood with big pores. Maple, birch, beech, and other diffuse-porous hardwoods in general show no definite relationship between the width of rings and density, except that usually narrow rings indicate brash wood.

Strength

Wood, when used in ordinary structures, is called upon to have three types of strength—tension, compression, and shear.

Tension—Tension is the technical term for a pulling stress. For example, if two men are having a tug of war with a rope, the rope is in tension. The tensile strength of wood, especially of the structural grades, is very high.

Compression— If, however, the men at opposite ends of a 2 × 4 are trying to push each other over, the timber is in compression. Tension and compression represent, therefore, exactly opposite forces.

Shear— If two or three planks are placed one upon the other between two blocks, and a person were to stand in the middle, the planks would bend (Fig. 2–8). It will be noted that at the outer ends the boards tend to slip past each other.

If the planks were securely spiked through from top to bottom, the slipping would be in a great measure prevented and the boards would act more as one piece of wood. In very solid timber there is the same tendency for the various parts of the piece to slip past each other. This tendency is called *horizontal shear*. A defect, such as a check, which runs horizontally through a piece of a timber and tends to separate the upper from the lower part, is a weakness in shear.

It is well to analyze this matter a little further. Suppose that the planks were spiked through at the center of span only, i.e., halfway between the blocks. Such spikes would not increase the stiffness of the planks. It is clear, therefore, that there is no horizontal shear near the center of the span (Fig. 2–9), and that the shear increases as one approaches either end of the beam. This will explain why, as most carpenters have doubtless observed, steel stirrups are used in concrete beams (weak in shear), why there is usually none

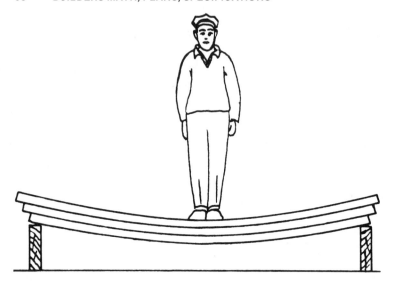

Fig. 2–8. Illustrating shear in lumber.

near the center, and why they are put closer and closer together near the ends of the beams.

For all practical purposes the compressive strength of wood may be considered to equal its tensile strength. It has been extremely difficult to make any direct measurements of the tensile strength of wood. In an experiment designed to ascertain the tensile strength of a specimen of wood, a 4″ × 4″ piece was selected. A portion about a foot in length near the center was carefully cut down on all four sides until it was exactly ¾-inch square. The test

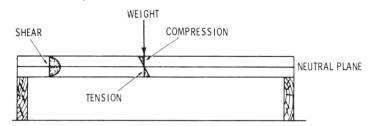

Fig. 2–9. Illustrating the different proportions of tension, compression, and shear.

specimen was placed in a machine which gripped the 4″ × 4″ ends securely and a pull was exerted. The specimen did not pull apart. The ¾-inch-square section held and actually pulled out of the end of the 4″ × 4″, leaving a ³/₄-inch square hole. This is an excellent illustration of how a piece may fail from shear rather than tension, the shear in this case being insufficient to prevent the ¾-inch-square piece from pulling out.

Deadwood

Because in some instances persons are prejudiced against the use of timber cut from dead trees, it is customary for individuals to specify that only timber cut from live trees will be accepted. It is true, however, that when sound trees that are dead are sawed into lumber and the weathered or charred outside is cut away, the resulting lumber cannot be distinguished from that coming from live trees except insofar as the lumber from dead trees may be somewhat seasoned at the time it is sawed. It must be remembered that the heartwood of a living tree is fully matured and that in the sapwood only a small portion of the cells are in a living condition. As a consequence, most of the wood cut from trees is already dead, even when the tree itself is considered alive.

For structural purposes, it may be said that lumber cut from fire- or insect-killed trees is just as good as any other lumber unless the wood has been subjected to further decay or insect attack.

Virgin and Second Growth

Occasionally an order calls for lumber of either virgin growth or second growth. The terms, however, are without significance, as an individual cannot tell one type from the other when it is delivered.

The virgin growth, which is also called old growth or first growth, refers to timber that grows in the forest along with many other trees, and therefore has suffered the consequence of the fight for sunlight and moisture.

The second growth is considered as that timber which grows up with less of the competition for sunlight and moisture that characterizes first-growth timber.

Because of environment, the virgin growth is usually thought of as wood of slow-growing type, whereas the second growth is considered as of relatively rapid growth, evidenced by wider annual rings. In such hardwoods as ash, hickory, elm, and oak these wider annual rings are supposed to indicate stronger and tougher wood, whereas in the conifers such as pine and fir, this condition is supposed to result in a weaker and brasher wood. For this reason, where the strength and toughness are desired, the second growth is preferred among hardwoods, and virgin growth is desired in conifers. Because of the variety of conditions under which both virgin and second growth timbers grow, because virgin growth may have the characteristics of second growth, and because second growth may have the characteristics of virgin timber, it is advisable in judging the strength of wood to rely upon its density and rate of growth rather than upon its being either virgin or second growth.

Time of Cutting Timber

The time when timber is cut has very little to do with its durability or other desirable properties if, after it is cut, it is cared for properly. Timber cut in the late spring, however, or early summer is more likely to be attacked by insects and fungi. In addition, seasoning will proceed much more rapidly during the summer months and, therefore, will result in checking, unless the number is shaded from the intense sunlight. There is practically no difference in the moisture content in green lumber cut either during the summer or winter.

Air-Dried and Kiln-Dried Wood

There is a prevailing misapprehension that air-dried lumber is stronger or better than kiln-dried lumber. Exhaustive tests have conclusively shown that good kiln-drying and good air-drying have exactly the same results on the strength of the wood. Wood increases in strength with the elimination of moisture content. This may account for the claim that kiln-dried lumber is stronger than air-dried lumber. This has little significance because in use wood will come to practically the same moisture content whether it has been kiln-dried or air-dried.

The same kiln-drying process cannot be applied to all species of wood. Consequently it must be remembered that lack of certain strength properties in wood may be due to improper kiln-drying. Similar damage also may result from air seasoning under unsuitable conditions.

Sapwood versus Heartwood

The belief is common that in some species the heartwood is stronger than the sapwood and that the reverse is the case in such species as hickory and ash. Tests have shown conclusively that neither is the case, and that sapwood is not necessarily stronger than heartwood or heartwood stronger than sapwood, but that density rather than other factors makes the difference in strength. In trees that are mature, the sapwood is frequently weaker, whereas in young trees the sapwood may be stronger. Density, proportion of spring and summer wood, then must be the basis of consideration of strength rather than whether the wood is sapwood or heartwood. Under unfavorable conditions, the sapwood of most species is more subject to decay than the heartwood.

The cells in the heartwood of some species are filled with various oils, tannins, and other substances, called extratives, which make these timbers rot-resistant. There is practically no difference in the strength of heartwood and sapwood, if they weigh the same. In most species, only the sapwood can be readily impregnated with preservatives, a procedure used when the wood will be in contact with the ground.

Defects and Deterioration

The defects found in manufactured lumber have several causes:

1. Those found in the natural log.
   a. Shakes.
   b. Knots.
   c. Pitchpockets.
2. Those due to deterioration:
   a. Rot.
   b. Dote.

3. Those due to imperfect manufacture:
   a. Imperfect machining.
   b. Wane.
   c. Machine burn.
   d. Checks and splits from imperfect drying.

Heart shakes (Fig. 2–10) are radial cracks that are wider at the pith of the tree than at the outer end. This defect is more commonly found in old trees than in young vigorous saplings; it occurs frequently in hemlock.

A star shake resembles a wind shake but differs from it in that

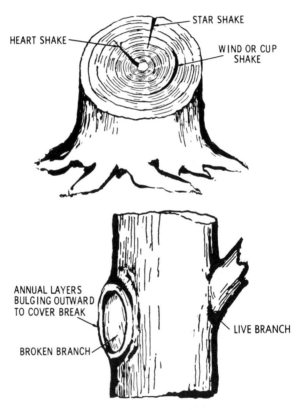

Fig. 2–10. **The cause of some lumber defects are found in the natural log.**

the crack extends across the center of the trunk without any appearance of decay at that point; it is larger at the outside of the tree. Heart and star shakes cause splits in lumber (Fig. 2–11B)

A wind or cup shake is a crack following the line of the porous part of the annual rings and is curved by a separation of the annual rings (Fig. 2–10). A wind shake may extend for a considerable distance up the trunk. Other explanations for wind shakes are expansion of the sapwood and wrenching from high wind (hence the name). Brown ash is especially susceptible to wind shake. Wind shakes cause cup checks in lumber (Fig. 2–11A).

Decay of Wood

Decay of lumber is the result of one cause, and one cause only, the work of certain low-order plants called fungi. All of these organisms require water, air, and temperatures well above freezing,

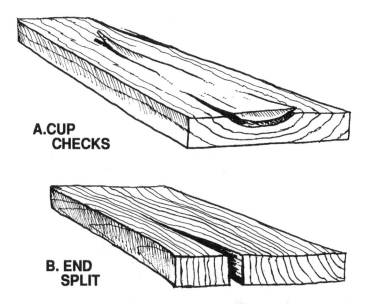

A. CUP CHECKS

B. END SPLIT

Fig. 2–11. Lumber defects caused by defects in the log. Cup checks are caused by wind shakes (A). End splits are caused by star and heart shakes (B). *(Courtesy Practical Restoration Reports)*

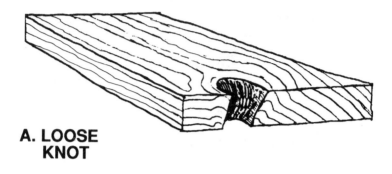

A. LOOSE
KNOT

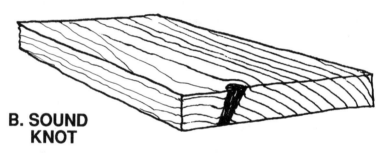

B. SOUND
KNOT

Fig. 2-12. Black or loose knots (A) are caused by broken branches in the log. Bulging layers of growth cause a large swirl of cross grain at the surface of the board surrounding the knot. Tight red knots (B) are caused by a live branch in the log. The knot will not fall out and grain is relatively straight around the knot. *(Courtesy Practical Restoration Reports)*

to live, grow, and multiply; consequently, wood that is kept dry, or that is dried quickly after wetting, will not decay.

Further, if it is kept continuously submerged in water even for long periods of time, it is not decayed significantly by the common decay fungi regardless of the wood species or the presence of sapwood. Bacteria and certain soft-rot fungi can attack submerged wood but the resulting deterioration is very slow. A large propor-

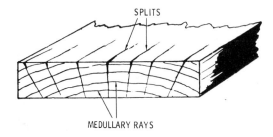

Fig. 2-13. A board with splits along the medullary rays. This condition is caused by too-rapid kiln drying.

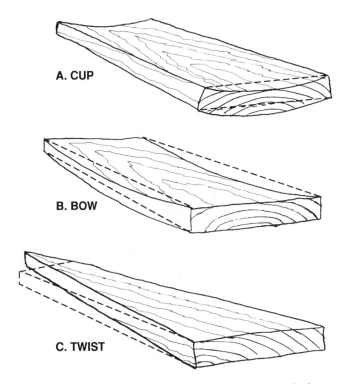

A. CUP

B. BOW

C. TWIST

Fig. 2-14. Warping is frequently caused by improper drying practices. Here the flatness of boards is distorted by irregular shrinkage. *(Courtesy Practical Restoration Reports)*

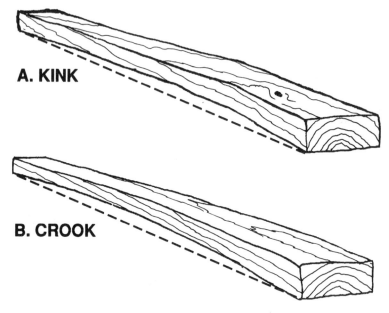

A. KINK

B. CROOK

Fig. 2-15. Warping can also affect straightness of a board. *(Courtesy Practical Restoration Reports)*

tion of wood in use is kept so dry at all times that it lasts indefinitely. Moisture and temperature, which vary greatly with local conditions, are the principal factors affecting rate of decay. When exposed to conditions that favor decay, wood deteriorates more rapidly in warm, humid areas than in cool or dry areas. High altitudes, as a rule, are less favorable to decay than low altitudes because the average temperatures are lower and the growing seasons for fungi, which cause decay, are shorter.

The heartwoods of some common native species of wood have varying degrees of natural decay resistance. Untreated sapwood of substantially all species has low resistance to decay and usually has a short service life under decay-producing conditions. The decay resistance of heartwood is greatly affected by differences in the preservative qualities of the wood extractives, the attacking fungus, and the conditions of exposure. Considerable difference in service life may be obtained from pieces of wood cut from the same species, or even from the same tree, and used under apparently similar conditions. There are further complications because, in a few species, such as the spruces and the true firs (not Douglas fir), heartwood and sapwood are so similar in color that they cannot be easily distinguished. Marketable sizes of some species such as southern pine and baldcypress are becoming largely second growth and contain a high percentage of sapwood.

Precise ratings of decay resistance of heartwood of different species are not possible because of differences within species and the variety of service conditions to which wood is exposed. However, broad groupings of many of the native species, based on service records, laboratory tests, and general experience, are helpful in choosing heartwood for use under conditions favorable to decay (Table 2-1). The extent of variations in decay resistance of individual trees or wood samples of a species is much greater for most of the more resistant species than for the slightly or nonresistant species.

Where decay hazards exist, heartwood of species in the resistant or very resistant category generally gives satisfactory service, but heartwood of species in the other two categories will usually require some form of preservative treatment. For mild decay conditions, a simple preservative treatment—such as a short soak in preservative after all cutting and boring operations are complete—will be adequate for wood low in decay resistance. For more severe decay hazards, pressure treatments are often required; even the very decay-resistant species may require preservative treatment for important structural or other uses where failure would endanger life or require expensive repairs.

Wood products sometimes are treated with preservative or fire-retarding salts, usually in water solution, to impart resistance to decay or fire. Such products generally are kiln dried after treatment. Mechanical properties are essentially unchanged by preservative treatment.

Properties are, however, affected to some extent by the combined effects of fire-retardant chemicals, treatment methods, and kiln drying. A variety of fire-retardant treatments have been stud-

Table 2-1. Grouping of Some Domestic Woods According to Heartwood Decay

Resistant or very resistant	Moderately resistant	Slightly or nonresistant
Baldcypress (old growth)[1]	Baldcypress (young growth)[1]	Alder
Catalpa	Douglas-fir	Ashes
Cedars	Honeylocust	Aspens
Cherry, black	Larch, western	Basswood
Chestnut	Oak, swamp chestnut	Beech
Cypress, Arizona	Pine, eastern white[1]	Birches
Junipers	Southern pine:	Buckeye
Locust, black[2]	Longleaf[1]	Butternut
Mesquite	Slash[1]	Cottonwood
Mulberry, red[2]	Tamarack	Elms
Oak:		Hackberry
Bur		Hemlocks
Chestnut		Hickories
Gambel		Magnolia
Oregon white		Maples
Post		Oak (red and
White		black species)
Osage orange[2]		Pines (other than
Redwood		long-leaf slash
Sassafras		and eastern
Walnut, black		white)
Yew, Pacific[2]		Poplars
		Spruces
		Sweetgum
		True firs (western
		and eastern)
		Willows
		Yellow poplar

[1]The southern and eastern pines and baldcypress are now largely second growth with a large proportion of sapwood. Consequently, substantial quantities of heartwood lumber of these species are not available.

[2]These woods have exceptionally high decay resistance.

ied. Collectively the studies indicate modulus of rupture, work to maximum load, and toughness are reduced by varying amounts depending on species and type of fire retardant. Work to maximum load and toughness are most affected, with reductions of as much as 45 percent. A reduction in modulus of rupture of as much as 20 percent has been observed; a design reduction of 10 percent is frequently used. Stiffness is not appreciably affected by fire-retardant treatments.

Wood is also sometimes impregnated with monomers, such as methyl methacrylate, which are subsequently polymerized. Many of the properties of the resulting composite are higher than those of the original wood, generally as a result of filling the void spaces in the wood structure with plastic. The polymerization process and both the chemical nature and quantity of monomers are variables that influence composite properties.

Nuclear Radiation

Very large doses of gamma rays or neutrons can cause substantial degradation of wood. In general, irradiation with gamma rays in doses up to about 1 megarad has little effect on the strength properties of wood. As dosage increases above 1 megarad, tensile strength parallel to grain and toughness decrease. At a dosage of 300 megarads, tensile strength is reduced about 90 percent. Gamma rays also affect compressive strength parallel to grain above 1 megarad, but strength losses with further dosage are less than for tensile strength. Only about one-third of the compressive strength is lost when the total dose is 300 megarads. Effects of gamma rays on bending and shear strength are intermediate between the effects on tensile and compressive strength.

Molding and Staining Fungi

Molding and staining fungi do not seriously affect most mechanical properties of wood because they feed upon substance within the structural cell wall rather than on the structural wall itself. Specific gravity may be reduced by from 1 to 2 percent, while most of the strength properties are reduced by a comparable or only slightly greater extent. Toughness or shock resistance, however, may be reduced by up to 30 percent. The duration of infection and the species of fungi involved are important factors in determining the extent of weakening.

Although molds and stains themselves often do not have a major effect on the strength of wood products, conditions that favor the development of these organisms are likewise ideal for the growth of wood-destroying (decay) fungi, which can greatly reduce mechanical properties.

Fungal Decay

Unlike the molding and staining fungi, the wood-destroying (decay) fungi seriously reduce strength. Even apparently sound wood adjacent to obviously decayed parts may contain hard-to-detect, early (incipient) decay that is decidedly weakening, especially in shock resistance.

All wood-destroying fungi do not affect wood in the same way. The fungi that cause an easily recognized pitting of the wood, for example, may be less injurious to strength than those that, in the early stages, give a slight discoloration of the wood as the only visible effect.

No method is known for estimating the amount of reduction in strength from the appearance of decayed wood. Therefore, when strength is an important consideration, the safe procedure is to discard every piece that contains even a small amount of decay. An exception may be pieces in which decay occurs in a knot but does not extend into the surrounding wood.

Blue Stain

In the sapwood of many species of both softwoods and hardwoods, there often develops a bluish-black discoloration known as blue stain. It does not indicate an early stage of decay, nor does it have any practicable effect on the strength of the wood. Blue stain is caused by a fungus growth in unseasoned lumber. Although objectionable where appearance is of importance, as in unpainted trim, blue stain need cause no concern for framing lumber. Precautions should be taken, however, to make sure that no decay fungus is present with the blue stain.

Weathering

Without protective treatment, freshly cut wood exposed to the weather changes materially in color. Other changes due to weathering include warping, loss of some surface fibers, and surface roughening and checking. The effects of weathering on wood may be desirable or undesirable, depending on the requirements for the particular wood product. The time required to reach the fully weathered appearance depends on the severity of the exposure to sun and rain. Once weathered, wood remains nearly unaltered in appearance.

The color of wood is affected very soon on exposure to weather. With continued exposure all woods turn gray; however, only the wood at or near the exposed surfaces is noticeably affected. This very thin gray layer is composed chiefly of partially degraded cellulose fibers and micro-organisms. Further weathering causes fibers to be lost from the surface but the process is so slow that only about ¼ inch is lost in a century.

In the weathering process, chemical degradation is influenced greatly by the wavelength of light. The most severe effects are produced by exposure to ultraviolet light. As cycles of wetting and drying take place, most woods develop physical changes such as checks or cracks that are easily visible. Moderate to low density woods acquire fewer checks than do high density woods. Vertical-grain boards check less than flat-grain boards.

As a result of weathering, boards tend to warp (particularly cup) and pull out their fastenings. The cupping tendency varies with the density, width, and thickness of a board. The greater the density and the greater the width in proportion to the thickness, the greater is the tendency to cup. Warping also is more pronounced in flat-grain boards than in vertical-grain boards. For best cup resistance, the width of a board should not exceed eight times its thickness.

Biological attack of a wood surface by micro-organisms is recognized as a contributing factor to color changes. When weathered wood has an unsightly dark gray and blotchy appearance, it is due to dark-colored fungal spores and mycelium on the wood surface. The formation of a clean, light gray, silvery sheen on weathered wood occurs most frequently where micro-organism growth is inhibited by a hot, arid climate or a salt atmosphere in coastal regions.

The contact of fasteners and other metallic products with the weathering wood surface is a source of color, often undesirable if a natural color is desired.

Insect Damage

Insect damage may occur in standing trees, logs, and unseasoned or seasoned lumber. Damage in the standing tree is difficult to control, but otherwise insect damage can be largely eliminated by proper control methods.

Insect holes are generally classified as pinholes, grub holes, and powderpost holes. The powderpost larvae, by their irregular burrows, may destroy most of the interior of a piece, while the surface shows only small holes, and the strength of the piece may be reduced virtually to zero.

No method is known for estimating the amount of reduction in strength from the appearance of insect-damaged wood, and, when strength is an important consideration, the safe procedure is to eliminate pieces containing insect holes.

Summary

Wood is the most versatile building material. Softwoods and hardwoods grow as trees with a fibrous cellular structure. Logs are converted into lumber and seasoned by drying. The physical characteristics of wood affect its performance and specific uses. Defects in lumber relate to defects in the logs it was cut from and how the lumber was handled during drying and storage. After wood is installed in a building, it can deteriorate by weathering, decay or insects unless protected.

Review Questions

1. Describe the physical structure of wood with a drawing. Label the parts.
2. What are the characteristics of softwoods and hardwoods? How is each type of wood used in building?
3. What is the difference between vertical-grain wood and flat-grain wood? How is each produced?
4. Why does wood have to be seasoned? How does the moisture content and size of wood change during seasoning?
5. How does shear affect the strength of wood?
6. What substances make some woods decay-resistant?
7. How do defects in logs relate to defects in lumber? Give some examples.
8. What causes decay in wood? How can you stop decay?
9. How does weathering affect the color of wood?
10. How long would it take for a 1-inch thick board to weather away by fiber loss?
11. Would vertical-grain or flat-grain boards hold up better on an outdoor deck? Why?

CHAPTER 3

Lumber and Wood Products

Lumber and Solid Wood

The basic construction material in carpentry is lumber. There are many kinds of lumber varying greatly in structural characteristics. Here, we deal with the lumber common to construction carpentry (Fig. 3–1).

Standard Sizes of Lumber

Lumber is usually sawed into standard lengths, widths, and thickness. This permits uniformity in planning structures and in ordering material (Table 3–1). Standards have been established for dimension differences between nominal size and the standard size. It is important that these dimension differences be taken into consideration when planning a structure. A good example of the dimension difference may be illustrated by the common 2 × 4. As may be seen in the table, the familiar quoted size (2 × 4) refers to a rough or nominal dimension, but the actual standard size to which the lumber is dressed is $1\frac{1}{2}'' \times 3\frac{1}{2}''$.

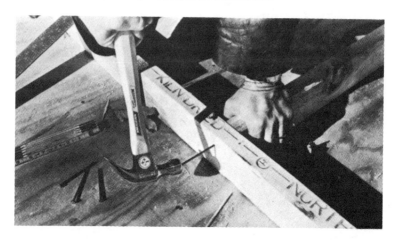

Fig. 3-1. Construction lumber. Note the classification "kiln dried" on one member. This stock will be used inside a structure. *(Courtesy of Vaughn & Bushnell)*

Table 3-1. Your Guide to New Sizes of Lumber

Standard (What You Order)	Nominal (What You Get)	
	* Dry or Seasoned	** Green or Unseasoned
1 × 4	$3/4 \times 3^1/2$	$25/32 \times 3^9/16$
1 × 6	$3/4 \times 5^1/2$	$25/32 \times 5^5/8$
1 × 8	$3/4 \times 7^1/4$	$25/32 \times 7^1/2$
1 × 10	$3/4 \times 9^1/4$	$25/32 \times 9^1/2$
1 × 12	$3/4 \times 11^1/4$	$25/32 \times 11^1/2$
2 × 4	$1^1/2 \times 3^1/2$	$1^9/16 \times 3^9/16$
2 × 6	$1^1/2 \times 5^1/2$	$1^9/16 \times 5^5/8$
2 × 8	$1^1/2 \times 7^1/4$	$1^9/16 \times 7^1/2$
2 × 10	$1^1/2 \times 9^1/4$	$1^9/16 \times 9^1/2$
2 × 12	$1^1/2 \times 11^1/4$	$1^9/16 \times 11^1/2$
4 × 4	$3^1/2 \times 3^1/2$	$3^9/16 \times 3^9/16$
4 × 6	$3^1/2 \times 5^1/2$	$3^9/16 \times 5^5/8$
4 × 8	$3^1/2 \times 7^1/4$	$3^9/16 \times 7^1/2$
4 × 10	$3^1/2 \times 9^1/4$	$3^9/16 \times 9^1/2$
4 × 12	$3^1/2 \times 11^1/4$	$3^9/16 \times 11^1/2$

*19% Moisture Content or under. **Over 19% Moisture Content.

Softwood Lumber Grades for Construction

The grading requirements of construction lumber are related specifically to the major construction uses intended and little or no further grading occurs once the piece leaves the sawmill. Construction lumber can be placed in three general categories—stress-graded, nonstress-graded, and appearance lumber. Stress-graded and nonstress-graded lumber are employed where the structural integrity of the piece is the primary requirement. Appearance lumber encompasses those lumber products in which appearance is of primary importance; structural integrity, while sometimes important, is a secondary feature.

Stress-Graded Lumber

Almost all softwood lumber nominally 2 to 4 inches thick is stress graded. Lumber of any species and size, as it is sawed from the log, is quite variable in its mechanical properties. Pieces may differ in strength by several hundred percent. For simplicity and economy in use, pieces of lumber of similar mechanical properties can be placed in a single class called a stress grade.

Visual grading is the oldest stress grading method. It is based on the premise that mechanical properties of lumber differ from mechanical properties of clear wood because of characteristics that can be seen and judged by eye. These visual characteristics are used to sort the lumber into stress grades (Table 3–2). The following are major visual sorting criteria:

1. Density
2. Decay
3. Heartwood and Sapwood
4. Slope of Grain
5. Knots
6. Shake
7. Checks and Splits
8. Wane
9. Pitch Pockets

Table 3-2. Visual Grades Described in the National Grading Rule[1]

Lumber classification	Grade name
Light framing (2 to 4 inches thick, 4 inches wide)[2]	Construction Standard Utility
Structural light framing (2 to 4 inches thick, 2 to 4 inches wide)	Select structural 1 2 3
Studs (2 to 4 inches thick, 2 to 4 inches wide)	Stud
Structural joists and planks (2 to 4 inches thick, 6 inches and wider)	Select structural 1 2 3
Appearance framing (2 to 4 inches thick, 2 to 4 inches wide)	Appearance

[1]Sizes shown are nominal.
[2]Widths narrower than 4 inches may have different strength ratio.

Nonstress-Graded Lumber

Traditionally, much of the lumber intended for general building purposes with little or no remanufacture has not been assigned allowable properties (stress graded). This category of lumber has been referred to as yard lumber; however, the assignment of allowable properties to an increasing number of former yard items has diluted the meaning of the term yard lumber.

In nonstress-graded structural lumber, the section properties (shape, size) of the pieces combine with the visual grade requirements to provide the degree of structural integrity intended. Typical nonstress-graded items include boards, lath, battens, crossarms, planks, and foundation stock.

Boards, sometimes referred to as "commons," are one of the more important nonstress-graded products. Common grades of boards are suitable for construction and general utility purposes. They are separated into three to five different grades depending upon the species and lumber manufacturing association involved. Grades may be described by number (No. 1, No. 2) or by descriptive terms (Construction, Standard).

Since there are differences in the inherent properties of the various species and in corresponding names, the grades for different species are not always interchangeable in use. First-grade boards are usually graded primarily for serviceability, but appearance is also considered. This grade is used for such purposes as siding, cornice, shelving, and paneling. Features such as knots and knotholes are permitted to be larger and more frequent as the grade level becomes lower. Second- and third-grade boards are often used together for such purposes as subfloors, roof and wall sheathing, and rough concrete work. Fourth-grade boards are not selected for appearance but for adequate strength. They are used for roof and wall sheathing, subfloor, and rough concrete form work.

Grading provisions for other nonstress-graded products vary by species, product, and grading association. Lath, for example, is available generally in two grades, No. 1 and No. 2; one grade of batten is listed in one grade rule and six in another. (For detailed descriptions consult the appropriate grade rule associations found in reference at the end of this volume.)

Appearance Lumber

Appearance lumber often is nonstress-graded but forms a separate category because of the distinct importance of appearance in the grading process. This category of construction lumber includes most lumber worked to a pattern. Secondary manufacture on these items is usually restricted to onsite fitting such as cutting to length and mitering. There is an increasing trend toward prefinishing many items. The appearance category of lumber includes trim, siding, flooring, ceiling, paneling, casing, base, stepping, and finish boards. Finish boards are commonly used for shelving and built-in cabinetwork.

Most appearance lumber grades are described by letters and combinations of letters (B&BTR, C&BTR, D). (See Standard Lumber Abbreviations in reference for definitions of letter grades.) Appearance grades are also often known as "Select" grades. Descriptive terms such as "prime" and "clear" are applied to a limited number of species. The specification FG (flat grain), VG (vertical grain), or MG (mixed grain) is offered as a purchase option for some appearance lumber products. In cedar and redwood, where there is a pronounced difference in color between heartwood and sapwood and heartwood has high natural resistance to decay, grades of heartwood are denoted as "heart." In some species and products two, or at most three, grades are available. A typical example is casing and base in the grades of C&BTR and D in some species and in B&BTR, C,C&BTR, and D in other species. Although several grades may be described in grade rules, often fewer are offered on the retail market.

Grade B&BTR allows a few small imperfections, mainly in the form of minor skips in manufacture, small checks or stains due to seasoning, and, depending on the species, small pitch areas, pin knots, or the like. Since appearance grades emphasize the quality of one face, the reverse side may be lower in quality. In construction, grade C&BTR is the grade combination most commonly available. It is used for high-quality interior and exterior trim, paneling, and cabinetwork, especially where these are to receive a natural finish. It is the principal grade used for flooring in homes, offices, and public buildings. In industrial uses it meets the special requirements for large-sized, practically clear stock.

The number and size of imperfections permitted increases as the grades drop from B&BTR to D and E. Appearance grades are not uniform across species and products, however, and official grade rules must be used for detailed reference. C is used for many of the same purposes as B&BTR, often where the best paint finish is desired. Grade D allows larger and more numerous surface imperfections that do not detract from the appearance of the finish when painted. Grade D is used in finish construction for many of the same uses as C. It is also adaptable to industrial uses requiring short-length clear lumber.

Select Lumber—Select lumber is of good appearance and finished or dressed. See Table 3–3 for grade names, descriptions, and uses.

Common Lumber—Common lumber is suitable for general construction and utility purposes and is identified by the grade names shown in Table 3–3.

Table 3-3. Lumber Grades

Grade	Description	Use
Select:		
A	High quality, practically clear.	Suitable for natural finishes.
B	High quality, generally clear with a few minor defects.	Suitable for natural finishes.
C	Several minor defects.	Adapted to high-quality paint finish.
D	A few major defects.	Suitable for paint finishes.
Common:		
No. 1	Sound and tight-knotted.	Use without waste.
No. 2	Less restricted in quality than No. 1.	Framing, sheathing, structural forms where strain or stress not excessive.
No. 3	Permits some waste with defects larger than in No. 2	Footings, guardrails, rough subflooring.
No. 4	Permits waste, low quality, with decay and holes.	Sheathing, subfloors, roof boards in the cheaper types of construction.

Plywood

Plywood is a glued wood panel made up of relatively thin layers, or plies, with the grain of adjacent layers at an angle, usually 90°. The usual constructions have an odd number of plies. The outside plies are called *faces* or *face* and *back* plies, the inner plies are called *cores* or *centers*, and the plies immediately below the face and back are called *crossbands*. The core may be veneer, lumber, or particleboard. The plies may vary as to number, thickness, species, and grade of wood.

As compared with solid wood, the chief advantages of plywood are its having properties along the length nearly equal to properties along the width of the panel, its greater resistance to splitting, and its form, which permits many useful applications where large sheets are desirable. Use of plywood may result in improved utilization of wood, because it covers large areas with a minimum amount of wood fiber. This is because it is permissible to use plywood thinner than sawn lumber in some applications.

The properties of plywood depend on the quality of the different layers of veneer, the order of layer placement in the panel, the glue used, and the control of gluing conditions in the gluing process. The grade of the panel depends upon the quality of the veneers used, particularly of the face and back. The type of the panel depends upon the glue joint, particularly its water resistance. Generally, face veneers with figured grain that are used in panels where appearance is important have numerous short, or otherwise deformed, wood fibers. These may significantly reduce strength and stiffness of the panels. On the other hand, face veneers and other plies may contain certain sizes and distributions of knots, splits, or growth characteristics that have no undesirable effects on strength properties for specific uses. Such uses include structural applications such as sheathing for walls, roofs, or floors.

Types of Plywood

Broadly speaking, two classes of plywood are available—hardwood and softwood. In general, softwood plywood is intended for construction use and hardwood plywood for uses where appearance is important.

Originally, most softwood plywood was made of Douglas fir, but western hemlock, larch, white fir, ponderosa pine, redwood, southern pine, and other species are now used.

Most softwood plywood used in the United States is produced domestically, and U.S. manufacturers export some material. Generally speaking, the bulk of softwood plywood is used where strength, stiffness, and construction convenience are more important than appearance. Some grades of softwood plywood are made with faces selected primarily for appearance and are used either with clear natural finishes or with pigmented finishes.

Hardwood plywood is made of many different species, both in the United States and overseas. Well over half of all hardwood panels used in the United States is imported. Hardwood plywood is normally used where appearance is more important than strength. Most of the production is intended for interior or protected uses,

although a very small proportion is made with glues suitable for exterior service. A significant portion of all hardwood plywood is available completely finished.

Plywood of thin, crossbanded veneers is very resistant to splitting and therefore nails and screws can be placed close together and close to the edges of panels.

Grades of Plywood

Plywood is graded according to defects on each surface and the type of glue used. Exterior or Interior plywood refers to the type of glue used to bond plies. Each face of the plywood has a letter grade—A, B, C, or D. Grade A means the face has no defects—it's perfect. Grade B means there are some defects; perhaps an area has a small patch. Grade C allows checks (splits) and small knotholes. Grade D allows large knotholes.

Fig. 3-2. **The multi-layer construction of plywood gives large tl panels greater strength and stability than solid wood.** *(Courtesy The American Plywood Assn.)*

The commonly available grades are AC Interior, AD Interior, AC Exterior, and CDX, which is used for sheathing.

Particle Board

The group of materials generally classified as "wood-base fiber and particle panel materials" includes such familiar products as insulation boards, hardboards, particleboards, and laminated paperboards. In some instances they are known by such proprietary names as "Masonite," "Celotex," "Insulite," and "Beaver board" or, in the instance of particleboards, by the kind of particle used such as flakeboard, chipboard, or oriented strand board.

These panel materials are all reconstituted wood (or some other lignocellulose like bagasse) in that the wood is first reduced to small fractions and then put back together by special forms of manufacture into panels of relatively large size and moderate thickness. These board or panel materials in final form retain some of the properties of the original wood but, because of the manufacturing methods, gain new and different properties from those of the wood. Because they are manufactured, they can be and are "tailored" to satisfy a use-need, or a group of needs.

Generally speaking, the wood-base panel materials are manufactured either (1) by converting wood substance essentially to fibers and then interfelting them together again into the panel material classed as building fiberboard, or (2) by strictly mechanical means of cutting or breaking wood into small discrete particles and then, with a synthetic resin adhesive or other suitable binder, bonding them together again in the presence of heat and pressure. These latter products are appropriately called particleboards.

Building fiberboards, then, are made essentially of fiberlike components of wood that are interfelted together in the reconstitution and are characterized by a bond produced by that interfelting. They are frequently classified as fibrous-felted board products. At certain densities under controlled conditions of hot-pressing, rebonding of the lignin effects a further bond in the panel product produced. Binding agents and other materials may be added during manufacture to increase strength, resistance to fire, moisture, or decay, or to improve some other property. Among the materials

added are rosin, alum, asphalt, paraffin, synthetic and natural resins, preservative and fire-resistant chemicals, and drying oils. Particleboards are manufactured from small components of wood that are glued together with a thermosetting synthetic resin or equivalent binder. Wax sizing is added to all commercially produced particleboard to improve water resistance. Other additives may be introduced during manufacture to improve some property or provide added resistance to fire, insects such as termites, or decay. Particleboard is among the newest of the wood-base panel materials. It has become a successful and economical panel product because of the availability and economy of thermosetting synthetic resins, which permit blends of wood particles and the synthetic resin to be consolidated and the resin set (cured) in a press that is heated.

Thermosetting resins used are primarily urea-formaldehyde and phenol-formaldehyde. Urea-formaldehyde is lowest is cost and is the binder used in greatest quantity for particleboard intended for interior or other nonsevere exposures. Where moderate water or heat resistance is required, melamine-urea-formaldehyde resin blends are being used. For severe exposures like exteriors or where some heat resistance is required, phenolics are generally used.

The kinds of wood particles used in the manufacture of particleboard range from specially cut flakes an inch or more in length (parallel to the grain of the wood) and only a few hundredths of an inch thick to fine particles approaching fibers or flour in size. The synthetic resin solids are usually between 5 and 10 percent by weight of the dry wood furnish. These resins are set by heat as the wood particle-resin blend is compressed in flat-platen presses.

As floor underlayment, particleboard provides (1) the leveling, (2) the thickness of construction required to bring the final floor to elevation, and (3) the indentation-resistant smooth surface necessary as the base for resilient finish floors of linoleum, rubber, vinyl tile and sheet material. Particleboard for this use is produced in 4- by 8-foot panels commonly $\frac{1}{4}$, $\frac{3}{8}$, or $\frac{5}{8}$ inch thick. Separate use specifications cover particleboard floor underlayment. In addition, all manufacturers of particleboard floor underlayment provide individual application instructions and guarantees because of the importance of proper application and the interaction effects of joists, subfloor, underlayment, adhesives, and finish flooring. Particleboard underlayment is sold under a certified quality program where established grade marks clearly identify the use, quality, grade, and originating mill.

Other uses for particleboard have special requirements, as for phenol-formaldehyde, a more durable adhesive, in the board. Particleboard for siding, combined siding-sheathing, and use as soffit linings and ceilings for carports, porches, and the like requires this more durable adhesive. For these uses, type 2 medium-density board is required. In addition, such agencies as Federal Housing Administration have established requirements for particleboard for such use. The satisfactory performance of particleboard in exterior exposure depends not only on the manufacture and kind of adhesive used, but on the protection afforded by finish. Manufacturers recognize the importance by providing both paint-primed panels and those completely finished with liquid paint systems or factory-applied plastic films.

Lumber Distribution

Large primary manufacturers and wholesale organizations set up distribution yards in lumber-consuming areas to more effectively distribute both hardwood and softwood products. Retail yards draw inventory from distribution yards and, in wood-producing areas, from local lumber producers.

Retail Yard Inventory

The small retail yards throughout the United States carry softwoods required for ordinary construction purposes and often small stocks of one or two hardwoods in the grades suitable for finishing or cabinetwork. Special orders must be made for other hardwoods. Trim items such as moulding in either softwood or hardwood are available cut to size and standard pattern. Cabinets are usually made by millwork plants ready for installation and many common styles and sizes are carried or cataloged by the modern retail yard. Hardwood flooring is available to the buyer only in standard patterns. Some retail yards may carry specialty stress grades of lumber such as structural light framing for truss rafter fabrication.

Some lumber grades and sizes serve a variety of construction needs. Some species or species groups are available at the retail level only in grade groups. Typical are house framing grades such as joist and plank which are often sold as No. 2 and Better (2&BTR). The percentage of each grade in a grouping is part of the purchase agreement between the primary lumber manufacturer and the wholesaler; however, this ratio may be altered at the retail level by sorting. Where grade grouping is the practice, a requirement for a specific grade such as No. 1 at the retail level will require sorting or special purchase. Grade grouping occurs for reasons of tradition and of efficiency in distribution.

Another important factor in retail yard inventory is that not all grades, sizes, and species described by the grade rules are produced and not all those produced are distributed uniformly to all marketing areas. Regional consumer interest, building code requirements, and transportation costs influence distribution patterns. Often small retail yards will stock only a limited number of species and grades. Large yards, on the other hand, may cater to particular construction industry needs and carry more dry dimension grades along with clears, finish, and decking. The effect of these variable retail practices is that the grades, sizes, and species outlined in the grade rules must be examined to determine what actually is available. A brief description of lumber products commonly carried by retail yards follows:

Stress-Graded Lumber for Construction—Dimension is the principal stress-graded lumber item available in a retail yard. It is primarily framing lumber for joists, rafters, and studs. Strength, stiffness, and uniformity of size are essential requirements. Dimension is stocked in all yards, frequently in only one or two of the general purpose construction woods such as pine, fir, hemlock, or spruce. 2 × 6, 2 × 8, and 2 × 10 dimension are found in grades of Select Structural, No. 1, No. 2, and No. 3; often in combinations of No. 2&BTR or possibly No. 3&BTR. In 2 × 4, the grades available would normally be Construction and Standard, sold as Standard and Better (STD&BTR), Utility and Better (UTIL&BTR), or Stud, in lengths of 10 feet and shorter.

Dimension is often found in nominal 2-, 4-, 6-, 8-, 10-, or 12-inch widths and 8- to 18-foot lengths in multiples of 2 feet. Dimension formed by structural end-jointing procedures may be found. Dimension thicker than 2 inches and longer than 18 feet is not available in large quantity.

Other stress-graded products generally present are posts and timbers, with some beams and stringers also possibly in stock. Typical stress grades in these products are Select Structural and No. 1 Structural in Douglas fir and No. 1SR and No. 2SR in southern pine.

Nonstress-Graded Lumber for Construction—Boards are the most common nonstress-graded general purpose construction lumber in the retail yard. Boards are stocked in one or more species, usually in nominal 1-inch thickness. Standard nominal widths are 2, 3, 4, 6, 8, 10, and 12 inches. Grades most generally available in retail yards are No. 1, No. 2, and No. 3 (or Construction, Standard, and Utility). These will often be combined in grade groups. Boards are sold square-ended, dressed and matched (tongued and grooved) or with a ship-lapped joint. Boards formed by endjointing of shorter sections may form an appreciable portion of the inventory.

Appearance Lumber—Completion of a construction project usually depends on a variety of lumber items available in finished or semi-finished form. The following items often may be stocked in only a few species, finishes, or in limited sizes depending on the yards.

Finish: Finish boards usually are available in a local yard in one or two species principally in grade C&BTR. Redwood and cedar have different grade designations. Grades such as Clear Heart, A, or B are used in cedar; Clear All Heart, Clear, and Select are typical redwood grades. Finish boards are usually a nominal 1 inch thick, dressed two sides to ¾ inch. The widths usually stocked are nominal 2 to 12 inches in even-numbered inches.

Siding: Siding, as the name implies, is intended specifically to cover exterior walls. Beveled siding is ordinarily stocked only in white pine, ponderosa pine, western red-cedar, cypress, or redwood. Drop siding, also known as rustic siding or barn siding, is usually stocked in the same species as beveled siding. Siding may be stocked as B&BTR or C&BTR except in cedar where Clear, A, and B may be available and redwood where Clear All Heart and Clear will be found. Vertical grain (VG) is sometimes a part of the grade designation. Drop siding sometimes is stocked also in sound knotted C and D grades of southern pine, Douglas fir, and hemlock. Drop siding may be dressed, matched, or shiplapped.

Flooring: Flooring is made chiefly from hardwoods such as oak and maple, and the harder softwood species, such as Douglas fir, western larch, and southern pine. Often at least one softwood and one hardwood are stocked. Flooring is usually nominal 1-inch thick dressed to $^{25}\!/_{32}$ inch, and 3- and 4-inch nominal width. Thicker flooring is available for heavy-duty floors both in hardwoods and softwoods. Thinner flooring is available in hardwoods, especially for recovering old floors. Vertical and flat grain (also called quartersawed and plainsawed) flooring is manufactured from both softwoods and hardwoods. Vertical-grained flooring shrinks and swells less than flat-grained flooring, is more uniform in texture, wears more uniformly, and the joints do not open as much.

Softwood flooring is usually available in B and Better grade, C Select, or D Select. The chief grades in maple are Clear No. 1 and No. 2. The grades in quartersawed oak are Clear and Select, and in plain-sawed Clear, Select, and No. 1 Common. Quartersawed hardwood flooring has the same advantages as vertical-grained softwood flooring. In addition, the silver or flaked grain of quartersawed flooring is frequently preferred to the figure of plain-sawed flooring. Beech, birch, and walnut and mahogany (for fancy parquet flooring) are also occasionally used.

Casing and base: Casing and base are standard items in the more important softwoods and are stocked by most yards in at least one species. The chief grade, B and Better, is designed to meet the requirements of interior trim for dwellings. Many casing and base patterns are dressed to $^{11}\!/_{16} \times 2\frac{1}{4}$; other sizes used include $^{9}\!/_{16} \times 3$, $3\frac{1}{4}$, and $3\frac{1}{2}$. Hardwoods for the same purposes, such as oak and birch, may be carried in stock in the retail yard or may be obtained on special order.

Shingles and shakes: Shingles usually available are sawn from western red cedar, northern white cedar, and redwood. The shingle grades are: Western red cedar, No. 1, No. 2, No. 3; northern white cedar, Extra, Clear, 2nd Clear, Clear Wall, Utility; redwood, No. 1, No. 2 VG, and No. 2 MG.

Shingles that are all heartwood give greater resistance to decay than do shingles that contain sapwood. Edge-grained shingles are less likely to warp than flat-grained shingles; thick-butted shingles less likely than thin shingles; and narrow shingles less likely than wide shingles. The standard thicknesses of shingles are described as ½, ⁵⁄₂, ¼, and ⁵⁄₂ (four shingles to 2 inches of butt thickness, five shingles to 2¼ inches of butt thickness, and five shingles to 2 inches of butt thickness). Lengths may be 16, 18, or 24 inches. Random widths and specified widths ("dimension" shingles) are available in western red cedar, redwood, and cypress.

Shingles are usually packed four bundles to the square. A square of shingles will cover 100 square feet of roof area when the shingles are applied at standard weather exposures.

Shakes are handsplit or handsplit and resawn from western red cedar. Shakes are of a single grade and must be 100 percent clear, graded from the split face in the case of handsplit and resawn material. Handsplit shakes are graded from the best face. Shakes must be 100 percent heartwood free of bark and sapwood. The standard thickness of shakes ranges from ⅜ to 1¼ inches. Lengths are 18 and 24 inches, and a 15-inch "Starter-Finish Course" length.

Important Purchase Considerations

The following outline lists some of the points to consider when ordering lumber or timbers.

1. *Quantity*—Feet, board measure, number of pieces of definite size and length. Consider that the board measure depends on the thickness and width nomenclature used and that the interpretation of this must be clearly delineated. In other words, nominal or actual, pattern size, etc., must be considered.

2. *Size*—Thickness in inches—nominal and actual if surfaced on faces. Width in inches—nominal and also actual if surfaced on edges. Length in feet—may be nominal average length, limiting length, or a single uniform length. Often a trade designation, "random" length, is used to denote a nonspecified assortment of lengths. Note that such an assortment should contain critical lengths as well as a range. The limits allowed in making the assortment random can be established at the time of purchase.

3. *Grade*—As indicated in grading rules of lumber manufacturing associations. Some grade combinations (B&BTR) are offi-

cial grades; other [Standard and Better (STD&BTR) light framing, for example] are grade combinations and subject to purchase agreement. A typical assortment is 75 percent Construction and 25 percent Standard, sold under the label STD&BTR. In softwood, each piece of such lumber typically is stamped with its grade, a name or number identifying the producing mill, the dryness at the time of surfacing, and a symbol identifying the inspection agency supervising the grading inspection. The grade designation stamped on a piece indicates the quality at the time the piece was graded. Subsequent exposure to unfavorable storage conditions, improper drying, or careless handling may cause the material to fall below its original grade.

Note that working or rerunning a graded product to a pattern may result in changing or invalidating the original grade. The purchase specification should be clear regarding regrading or acceptance of worked lumber. In softwood lumber, grades for dry lumber generally are determined after kiln drying and surfacing. This practice is not general for hardwood factory lumber, however, where the grade is generally based on grade and size prior to kiln drying.

4. *Species or groupings of wood*—Douglas fir, cypress, Hem-Fir, etc. Some species have been grouped for marketing convenience; others are traded under a variety of names. Be sure the species or species group is correctly and clearly depicted on the purchase specification.

5. *Product*—Flooring, siding, timbers, boards, etc. Nomenclature varies by species, region, and grading association. To be certain the nomenclature is correct for the product, refer to the grading rule by number and paragraph.

6. *Condition of seasoning*—Air dry, kiln dry, etc. Softwood lumber dried to 19 percent moisture content or less (S-DRY) is defined as dry by the American Lumber Standard. Other degrees of dryness are partially air dried (PAD), green (S-GRN), and 15 percent maximum (KD in southern pine). There are several specified levels of moisture content for redwood. If the moisture requirement is critical, the levels and determination of moisture content must be specified.

7. *Surfacing and working*—Rough (unplaned), dressed (surfaced), or patterned stock. Specify condition. If surfaced, indicate

S4S, S1S1E, etc. If patterned, list pattern number with reference to the appropriate grade rules.

8. *Grading rules*—Official grading agency name, product identification, paragraph number or page number or both, date of rules or official rule volume (rule No. 16, for example).

9. *Manufacturer*—Name of manufacturer or trade name of specific product or both. Most lumber products are sold without reference to a specific manufacturer. If proprietary names or quality features of a manufacturer are required, this must be stipulated clearly on the purchase agreement.

Summary

Lumber is the basic construction material used in construction. Wood is a highly variable product so lumber is standardized by sizes and grades. The three classes of grading are stress-graded, nonstress-graded, and appearance. Plywood is made up of layers of wood veneer in wide long panels that have several advantages over solid wood lumber. Particle-board is made of small chunks of wood compressed into a large panel. Understanding lumber and wood product distribution and inventories leads to more successful and economic construction projects.

Review Questions

1. Why is lumber standardized by size and grade?
2. What grades are used for structural light framing?
3. What grade of lumber would you use for kitchen cabinet construction?
4. What are the advantages of plywood over solid lumber?
5. What grades and species of select softwood lumber are available at your local retail lumber yard?

CHAPTER 4

Strength of Timbers

The various mechanical properties of woods have been investigated by exhaustive testing in many laboratories, most notably the Forest Products Laboratory of the U.S. Department of Agriculture at Madison, Wisconsin. In addition, much research has been done in civil and agricultural engineering laboratories at many state universities.

In the shop and in the field, the fitness of any species of wood for a given purpose depends on various properties. When treating the strength, stiffness, hardness, and other properties of wood, many technical terms are used. For an understanding of these terms, the following definitions are given.

Definitions

Bending forces—Forces that act on some members of a structural frame and tend to deform them by flexure.

Brittleness—Breaking easily and suddenly, usually with a comparatively smooth fracture; the opposite of *toughness*, sometimes incorrectly called *brashness*, which refers more to brittleness. Old and extremely dry wood is inclined to brashness; green or wet wood is tougher, though not as strong in most cases.

Compression—The effect of forces that tend to reduce the dimensions of a member, or to shorten it.

Deformation—A change of shape or dimension; disfigurement, such as the elongation of a structural member under tension.

Ductile—A term not applicable to wood; it is the property of a metal that allows it to be hammered thin or drawn into wires.

Elastic limit—The greatest stress that a substance can withstand and still recover completely when the force or strain is removed.

Factor of safety—The ratio between the stress at failure and allowable design stress. If the stress at failure in a bending beam is 4350 psi (pounds per square inch), and the timber is stress-graded at 1450f (allowable fiber stress in psi) the safety factor is 4350/1450, or 3.

Force—In common parlance, a *pull* or a *push;* that which would change the state of a body at rest, or would change the course of a body in motion (Newton's Law).

Load—Pressure acting on a surface, usually caused by the action of gravity.

Member—A part of a structure, such as a column, beam, or brace, which usually is subjected to compression, tension, shear or bending.

Modulus, or coefficient, of elasticity—Stress, either tension or compression, divided by the elongation or contraction per unit of length. Inside the elastic limit, the modulus of elasticity is approximately constant for most materials. In wood, it varies greatly with different species, with moisture content, and even in pieces sawed from the same log; in other words, it is subject to considerable natural variation.

Modulus of rupture—The calculated fixed stress in a beam at the point of rupture. Since the elastic limit will have been passed at this point, it is not a true fiber stress, but it is a definite quantity, and the personal factor is not involved when obtaining it.

Permanent set—When a member, either metal or wood, is stressed beyond its elastic limit, or subjected to stresses that may be far below its elastic limit for extremely long periods of time, it may take on permanent deformation. Permanent set in timbers does not mean nor imply that the timbers have been weakened.

Resilience—A synonym for *elasticity*. The property that enables a substance to spring back when a deforming force is removed.

Shear—The effect of forces, external or internal, which causes bodies or parts of bodies to slide past each other.

Strain—Alterations in the form of a member caused by forces acting on the member.

Strength—The power to resist forces, which may be tensile, compressive, or shearing, without breaking or yielding.

Stress—Distributed forces, such as pounds per square inch or tons per square foot. Within the elastic limit of materials, *stress* is approximately proportional to *strain*. This statement is called Hooke's Law.

Tenacity—A synonym for *tensile strength;* the power to resist tearing apart.

Tension—A force that tends to tear a body apart or elongate it.

Toughness—Strong but flexible; not brittle; nearly the same as *tenacity.*

Ultimate strength—The stress developed just before failure is evident.

Yield point—This property is not evident in timbers. In steels, it occurs after the elastic limit has been passed. In materials that show no defined yield point, it may be arbitrarily assumed or defined as the stress where a permanent set occurs.

As an example of the uses for these terms, the tie rod in a truss resists being pulled apart because of its *tensile strength.* The *stress* thus applied *strains* the rod, *deforming* or *elongating* it. It is *stretched* and a *contraction* of the area of its cross section results. If the *load* is not sufficient to *stress* the material past its *elastic limit,* the rod will return to its original length when the load is removed, depending on the duration of the load. If the *load* is heavy enough to stretch the rod past its *elastic limit,* it will not return to its original length when the *load* is removed, but it will remain *permanently set* if it is not pulled in two. The *elastic limit* is reached when *elongation* becomes proportionally greater than the *loading.* If the *load* is increased to the point where the rod breaks, or where its *tenacity* is overcome, it is said to be *ruptured.* The rod itself is called a *member* of the truss.

Tension

A tension test is made as indicated in Fig. 4–1. Although modern testing machines are by no means as simple as the apparatus shown, it serves well to show how such tests are made. The specimen is placed in the machine, gripped at each end, and the load is progressively increased until the material reaches its failure point. Usually the elongation is not of any significance unless it is desired to determine the modulus of elasticity. The allowable fiber stress indicated by modern stress-grading, such as 1200f or 1450f, apply equally to extreme fiber stress in bending and to tensile stresses.

Example—A truss member is subjected to a tensile force of 50,000 pounds. What size timber of No. 1 dense yellow pine (1600f grade) will be required?

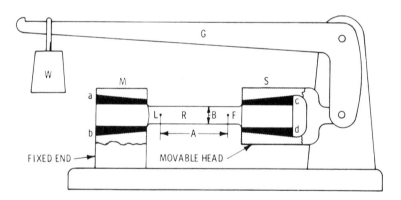

Fig. 4–1. The tension test. The specimen R is placed in the wedge grips a, b, c, and d, thus applying tension between the fixed end and the movable head of the machine. The movable head is connected to the scale lever G on which the weight W slides; this arrangement is similar to an ordinary weighing scale. Two center marks (L and F) are punched on the specimen at a standard distance (A) apart. When testing, the pull on the specimen is gradually increased by moving W to the left; dimensions A and B are then measured after each load increase.

This is the calculation:

$$\frac{5000}{1600} = 31.25 \text{ square inches cross section}$$

Therefore, either a standard-dressed 4″ × 10″ (3½″ × 9¼″) or a 6″ × 6″ (5½″ × 5½″) should be adequate.

Compression

A column supporting a load which tends to crush it is said to be in *compression.* Allowable compression parallel to the grain in stress-graded lumber is usually slightly less than the allowable bending and tensile stresses. For No. 1 dense timbers, 1600f grade, the compression stress is 1500 pounds per square inch, and for No. 2 dense, 1200f grade, it is 900 pounds per square inch. The builder need have no other concern than to see that the specified stresses are not exceeded. For loads applied across the grain, these stresses are much less; for the two grades mentioned above, the cross-grain stress is 455 pounds per square inch for both grades.

When making a compression test (Fig. 4-2), a prepared speci-

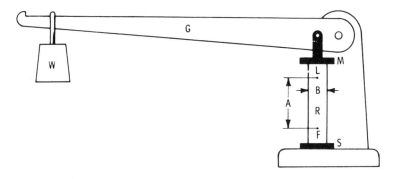

Fig. 4-2. The compression test. The specimen R is placed between two plates (M and S), and a compression stress of any desired intensity is applied by moving the weight W on the lever G. As the load is gradually increased, the changes in dimensions A and B are noted, and a final result can be obtained that will indicate the amount of compression which the specimen can withstand.

men is placed between two plates, and a measured load is applied; the load is increased progressively until the failure point is reached. Actually, modern wood-compression testing is not quite so simple. An instrument called a "compressometer" is pinned to the side of the specimen with sharp-pointed screws at points corresponding to points L and F in the illustration. It is fitted with a series of levers which are connected to a dial gauge, usually reading to .0001 inch. The stem of the dial gauge is depressed until it shows a positive reading, and compression in the specimen releases a part of the gauge. In this way, a sudden failure will release the gauge entirely, and jamming and ruining of the instrument is avoided. Usually at least one of the plates through which pressure is applied is a cast-iron hemisphere, thereby assuring an evenly distributed pressure over the entire cross section of the specimen.

Working Stresses for Columns

The amount of allowable loads on wood columns has been, and continues to be, a subject that is open to some discussion. The matter is complicated by the fact that columns of different lengths and diameters do not behave in the same manner under loadings. For the rather short column, failure will be caused by the actual crushing of the fibers of the wood, and the full compressive strength of the wood may thus be utilized. For a slightly longer column, failure may be caused by a sort of diagonal shearing action. For a long column, it will probably fail by bending sidewise and breaking. No one method can be adapted exactly for calculating allowable loads on columns of all lengths and slenderness ratios. Column formulas are numerous. Some are based on empirical data, or the results of actual testing. Only one, the Euler (pronounced *oiler*) formula seems to be based on purely mathematical calculations. It is of German origin, and, in its original form, it is so cumbersome that few designers in the U.S. care to use it. It assumes that a column will fail by *bending and breaking*. This is assured only if the column is long and comparatively slender; however, a *modified* Euler formula is greatly favored by present-day timber designers. It is written as follows:

Allowable load per square inch of cross section $= \dfrac{.3E}{(l/d)^2}$ where,

E = modulus of elasticity of the timber used,
l = unsupported length of the column, in inches,
d = least side, or diameter of a round column.

This formula is applicable only when the results of its use do not indicate a higher stress than the maximum allowable unit stresses as defined by the stress grade of the timber used. The ratio l/d in the above equation is sometimes called the slenderness ratio. As an example of the use of this formula, take the following problem:

How much of a load may safely be imposed on a $6'' \times 6''$ dressed yellow pine column which is 12 feet long? The slenderness ratio, or l/d, of $6'' \times 6''$ columns 12 feet long is $144/6 = 24$, and the modulus of elasticity of almost all good yellow pine is 1,760,000 pounds per square inch. This is the calculation:

$$\frac{.3 \times 1,760,000}{24 \times 24} = 917 \text{ pounds per square inch}$$

The cross-sectional area of standard $6'' \times 6''$ timbers is 30.25 square inches. Therefore, the total load allowable on the column will then be:

$$917 \times 30.25 = 27,739 \text{ pounds}$$

An important point with respect to timber in compression is that the ends should be cut exactly square so that there will be a full bearing surface; otherwise the timber will be subjected at the ends to more than the working stress (Fig. 4–3).

Shearing Stresses

Shearing stresses in wood are dangerous only in the direction parallel to the grain. It is almost impossible to shear the material across the grain until the specimen has been *crushed*. The crushing

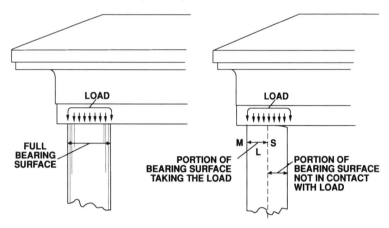

Fig. 4-3. Good and poor column bearing surfaces illustrate the importance of squaring columns accurately when cutting them. The entire top of the column must be in contact with the load member so that the pressure per square inch of cross section on the column will correspond to the allowable working pressure for which the column was designed. If the portion MS of the bearing surface in contact is only $\frac{1}{2}$ the entire surface, then the stress applied on the top of the column will be twice that of full contact, as shown by LF.

strength, then, and not the shearing resistance of the wood, will govern the maximum stress that can be applied to the wood.

The standard shearing test for woods with the grain was developed by the American Society for Testing Materials, and the procedure is standardized. The specimens are standard dimensions, shaped as shown in Fig. 4-4, in pairs, one with the notch at right angles to that in the other. The results of the tests on the two specimens are averaged, but usually they vary only slightly. The blocks are tested in a special shear tool, which is loaded in a universal testing machine, with the load applied at a rate of .024 inch per minute.

The results of testing different types of woods vary widely. To provide for this lack of uniformity, the shear allowances in stress-rated timbers may contain a reduction factor (safety factor) of as much as 10, or it may be as little as 2 or 3. This is necessary, because a piece which is below average in shearing resistance may appear anywhere.

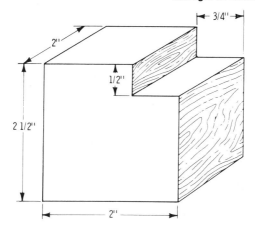

Fig. 4-4. The A.S.T.M. standard wood-shear test specimen.

The distinctions *across the grain* and *with the grain* should be carefully noted. Wet or green wood, in general, shears approximately one-fifth to one-half as easily as dry wood; a surface parallel to the rings (tangent) shears more easily than one parallel to the medullary rays. The lighter conifers and hardwoods offer less resistance than the heavier kinds, but the best pine shears one-third to one-half more readily than oak or hickory, thereby indicating that great shearing strength is characteristic of "tough" woods.

Horizontal Shears

The types of shears discussed in the preceding section are called *external* shears because they are caused by forces that originate outside the body of the material, and for the most part they are evident and readily provided for. In addition, in every loaded beam, a system of *internal* stresses are set up. This may be explained by observing that every shearing force results from *two* forces which are *unbalanced*. They do not meet at the same point, and a *stress couple* is set up which can be met and held motionless only by another couple which acts in opposition to it. This is illustrated in Fig. 4-6; it represents a "particle" of indefinite size, but possibly infinitely small, that has been extracted from the body of a

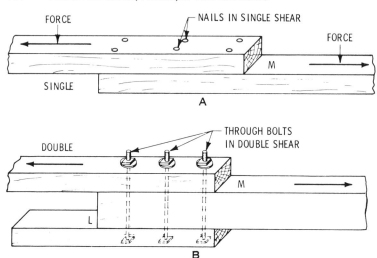

Fig. 4-5. Examples of single and double shear. In A, the nails are in single shear. This is the assumption made as a general rule when calculating the strength of nailed joints, but nails may be placed in double shear, similar to the bolts in B, if they are long enough to almost penetrate all three members; the nails in double shear will safely carry twice the load which could safely be placed on nails in single shear. However, most nailed joints are designed as if the nails were in single shear, though they may actually be in double shear. In B, the bolts are truly in double shear, but the joint is not twice as strong as a plain lapped joint if the center member is not at least as thick as the combined thicknesses of the outer members. The outer members are usually both the same thickness.

beam which has been stressed by bending. In Fig. 4–6A, the particle has been subjected to a vertical shearing force, or pair of forces, since there can be no action without an opposing and equal reaction. The particle is *unstable;* since the two forces do not meet at a common point, the particle tends to revolve in a counterclockwise direction. In Fig. 4–6B, a pair of horizontal forces is supplied; these forces represent another stress couple that balances or neutralizes the vertical shearing forces and are known as the *internal horizontal shears.* A reasonable deduction, therefore, is that at any point on the beam, there exist internal horizontal shears equal in intensity to

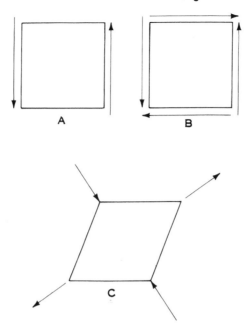

Fig. 4-6. The actions of vertical and horizontal shears.

the external vertical shears. It is the horizontal shears which are dangerous in wood beams, because timbers have a low resistance to shearing with the grain.

Fig. 4–6C shows how the four forces may be resolved into a single pair of concurrent forces which *do* meet at a common point. These are not shearing forces but are the result of shearing forces. When the material has approximately the same resistance to shears in all directions, such as concrete or steel, this is what occurs. Each particle tends to elongate in one diagonal and forms the "diagonal tension" which is so dangerous in reinforced concrete beams. In fact, it is the most dangerous stress in such beams, and to resist it, elaborate web reinforcing and bent-up bars are provided at points where this stress is highest, usually near the ends of the beams. In I beams, diagonal tension results in buckling or wrinkling of the relatively thin webs. In wood beams, the forces are not resolved in this way, since wood is strong enough to resist the vertical components, but the horizontal components tend to split the beam, usually at or near the ends near the center of the height. Wood beams which are season checked at the ends, as many are, are low in resistance to such stresses.

If the depth of a wood beam is greater than one-tenth to one-twelfth of its span, horizontal shears, and not bending strength, often govern its ability to carry loads. Shears are usually not dangerous in wood beams unless they are relatively deep and heavily loaded.

Transverse or Bending Stress

This is the kind of stress present on numerous building timbers, such as girders, joists, or rafters, that causes a deflection or bending between the points of support. What takes place when these or similar members are subjected to bending stress is considered under the subject of *beams* in this chapter.

Stiffness

By definition, stiffness is that quality possessed by a beam or other timber to resist the action of a bending force. The action of the bending force tends to change a beam from a straight to a curved form; that is, a *deflection* takes place. When a load is applied, the beam originally assumed to be straight and horizontal sags or bends downward between the supports. The amount of downward movement measured at a point midway between the supports is the amount of deflection. The action of beams subjected to bending forces is described as follows:

If a load of 100 pounds placed in the middle of a stick which is 2″ × 2″ and 4 feet long, supported at both ends, bends or deflects this stick one-eighth of an inch (in the middle), then 200 pounds will bend it about one-fourth inch, 300 pounds, three-eighths inch, etc., the deflection varying directly as the load. This is in accordance with Hooke's Law, which states that stress is proportional to strain. Soon, however, a point is reached where an additional 100 pounds adds more than one-eighth inch to the deflection—the limit of elasticity has been exceeded. Taking another piece from the straight-grained and perfectly clear plank of the same depth and width but 8 feet long, the load of 100 pounds will deflect it by approximately 1 inch. Doubling the length reduces the stiffness eightfold. Stiffness, then, decreases as the cube of the length.

If AB in Fig. 4–7 is a piece of wood, and D is the deflection produced by a weight or load, then

$$\text{deflection } (D) = \frac{Pl^3}{48EI}$$

where,

> P = the load, concentrated at the center of the span, in pounds,
> l = the length of the span, in feet,
> E = the modulus of elasticity of the material,
> I = the moment of inertia (for rectangular beams $= \frac{bd^3}{12}$
> where b = width and d = height).

The following rules, in conjunction with Figs. 4–8, 4–9, and 4–10, define the stiffness and strength of practically all types of wood beams.

1. For beams with a rectangular cross section, equal depths, and equal spans, their load-carrying capacity varies *directly* as their widths.
2. For beams with equal widths and equal spans, their strength varies directly as the square of their depth.

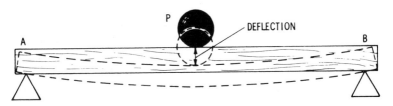

Fig. 4–7. A simple beam, loaded at the middle and supported at both ends, is used to illustrate the term "deflection."

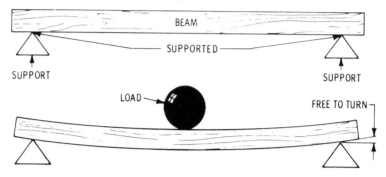

Fig. 4-8. A beam supported at the ends. The ends of the beam are free to follow any deflection, thus offering no resistance and rendering the beam less stiff than when the ends are fixed.

3. If depths and widths are the same, strengths vary *inversely* as the lengths of the spans.
4. Their stiffness, or resistance to deflection, will vary *inversely* as the *cubes* of their spans, other factors being equal.
5. Their stiffness will vary directly as the *cubes* of their depths, other factors being equal.
6. Other factors being equal, stiffness will vary *directly* as their widths.

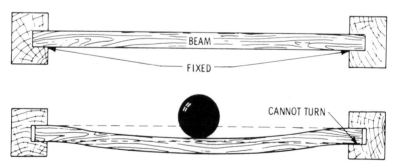

Fig. 4-9. A beam supported at the ends. The ends are gripped or embedded in some unyielding substance so that they cannot turn or follow the deflection of the beam under an applied load. The beam then deflects in a compound curve, thus adding extensively to its stiffness. Therefore, a beam with fixed ends will deflect less than one with supported ends.

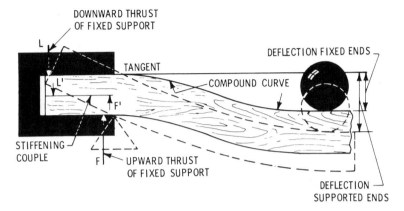

Fig. 4-10. One end of a beam illustrating the stiffening effect of fixed ends as compared with supported ends. When the ends are fixed, the deflection of the beam will be resisted by an upward thrust, indicated by F, and a downward thrust, L. These thrusts form a resisting or stiffening couple which holds the portion of the beam embedded in the bearing in a horizontal position, thereby causing the beam to deflect in a compound curve, which increases its stiffness. The dotted lines show the excess deflection for the same load if the beam were simply supported at the ends.

7. If a beam is split horizontally, and the two halves are laid side by side, they will carry only *one-half* as much loading as the original beam.

These relations are not strictly true for I beams because of their irregular shapes, but they are approximately true for all types of beams. It is usually most economical with materials to use as deep a beam as can conveniently be employed. Note that double 2″ × 4″ trimmers over window or door heads, if set edge up, are 8 times as strong and 32 times as stiff as when placed flatwise.

Both strength and stiffness are greater in dry timber than in green or wet wood of the same species. A piece of long-leaf yellow pine is 30% to 50% stronger and 30% stiffer when in an air-dry condition than when green. In general, both strength and stiffness are proportional to densities, or dry weights, although this is not invariably true. Edge-grain pieces are usually stronger and stiffer than those in which the tangent to the rings runs horizontally, but not appreciably so. There is little or no difference in the sapwood and heartwood of the same species, if the densities are the same. The tool handle of red heartwood is as serviceable as the handle of white sapwood, although white sapwood handles are still called "premium grade."

Modulus of Elasticity

Since it is desirable, and for many purposes essential, to know beforehand that a given piece with a given load will bend only by a given amount, the stiffness of wood is usually stated in a uniform manner under the term "modulus" (measure) of elasticity. For good grades of Douglas fir and yellow pine that are stress-rated, the modulus of elasticity is 1,760,000 pounds per square inch.

Beams

A beam is a single structural member, usually horizontal or nearly so, which carries a load or loads over a given space. At their supports, beams may be:

1. Freely supported merely means the beams are resting on their bearings.
2. Restrained, or partially fixed at their bearings. Although some designers choose to consider such restraint in their designs, the actual degree of restraint can never be accurately determined, so restrained beams are more often considered as being freely supported.
3. Fixed at their supports. In wood beams, this condition is rarely found; in steel frames, it is not unknown; and in reinforced concrete frames, it is quite common. Attempts to fix the ends of wood beams are rarely permanent. Building the ends of a beam, *any* beam, into a wall or casting it into concrete for a short distance does not fix the beams at their supports.

Allowable Loads on Wood Beams

The allowable loads on freely supported wood beams of any species are readily calculated if the timbers are stress rated and the

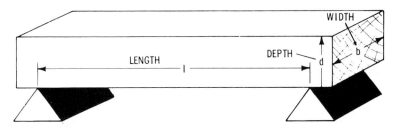

Fig. 4-11. A beam resting on knife-edge supports illustrates the terms length, width, and depth.

allowable fiber-stress is known. For beams with a loading which is evenly distributed along its span, this is the equation to use:

$$W = \frac{f \times b \times d^2}{9 \times L}$$

where,

$\quad W$ = allowable evenly distributed loading, in pounds,
$\quad f$ = allowable fiber stress in pounds per square inch,
$\quad b$ = width of the beam, in inches,
$\quad d$ = depth of the beam, in inches,
$\quad L$ = length of the span, in feet.

Fig. 4-11 defines the dimensions of a beam—b, width; d, depth; and L, length.

$\quad$*Problem*—What will be the maximum allowable load on a beam whose nominal size is 6″ × 12″ and whose actual size is 5½″ × 11¹/₂″, with an 18 foot span? The timber is to be 1500f stress rated. This is the calculation:

$$\frac{1500 \times 5.5 \times 11.5 \times 11.5}{9 \times 18} = 6735 \text{ pounds}$$

$\quad$If the loading is to be concentrated at the center of the span, use one-half the load as calculated by the formula given; therefore, for the same timber as calculated, with the same span, the allowable concentrated load will be

$$\frac{6735}{2} = 3367 \text{ pounds}$$

Breaking Loads on Wood Beams

Breaking loads are of no interest to the builder because the term is meaningless unless some explanation is made, since wood is extremely sensitive to the *duration* of loads. The load that would break a beam over a long period of time, for example, 10 years, will be only approximately $\frac{9}{16}$ of the load that would break it in a few minutes. The stresses specified in stress-rating lumber and timbers recognize this phenomenon. It is presumed that the full design loading will not be applied for more than 10 years during the life of the structure, and the time may be either cumulatively intermittent or continuous. It is also presumed that 90% of the full design loading may safely be applied for the full life of the structure. These presumptions make the use of stress-rated lumber quite conservative.

Example—What is the safe working load, concentrated at the center of the span, for a full-size 6″ × 10″ white oak timber with a 12-foot span, stress-rated 1900f, if it is laid flatwise (Fig. 4–12A)? If it is set edge up (Fig. 4–12B)?

This is the calculation for the timber laid flatwise:

$$\frac{1900 \times 10 \times 6 \times 6}{9 \times 12} = 6333$$

$$\frac{6333}{2} = 3167 \text{ pounds}$$

For the timber set edge up:

$$\frac{1900 \times 6 \times 10 \times 10}{9 \times 12} = 10,555$$

$$\frac{10,555}{2} = 5278 \text{ pounds}$$

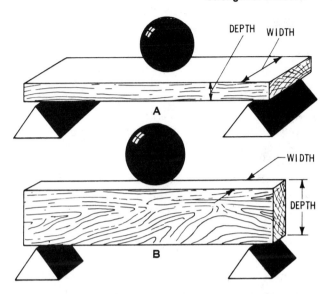

Fig. 4-12. The terms "width" and "depth" of a beam depend on the position of the beam. In A, the broad side is the width of the beam, whereas in B, which is the same beam turned over 90°, the narrow side is the width.

Distributed Load

Instead of placing all the load in the middle of a beam, as in the example just given, the load may be regarded as being distributed; that is, the beam is uniformly loaded (Fig. 4-13). Although this

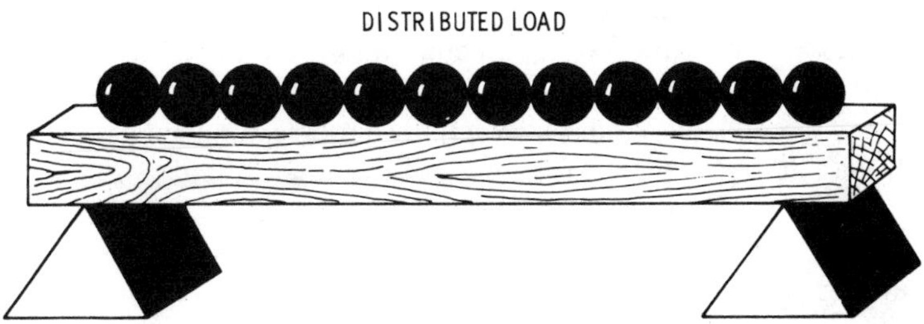

DISTRIBUTED LOAD

Fig. 4-13. A distributed load is indicated by the iron balls equally spaced along the beam between the supports.

type of loading is actually a series of concentrated loads, it is often found in actual practice, and it is usually considered as being *uniformly* distributed.

Cantilever Beams

A cantilever beam (Figs. 4-14, 4-15) is firmly fixed at one end, or it is freely supported but with the end running back some distance to a support above it. The loading may be either distributed or concentrated at any point on the span. All beams which project beyond a support and which carry a load at the free end are classed as cantilevers. Stresses set up in the beam external to the outside support are the same as when the beam is rigidly fixed at the support. Fig. 4-16 illustrates the comparison between the working loads of variously supported beams, contrasting middle and distributed loads of beams supported at both ends with equal loads on cantilever beams. In the illustration, $P =$ the concentrated loads, $W =$ the evenly distributed load, $f =$ the fiber stress, $b =$ the width of the beam, $d =$ the depth of the beam, and $L =$ the length of the span.

Wind Loads on Roofs

Snow on roofs, if the roofs are of a slope that is low enough to permit the snow to lie in place, will weigh approximately 8 pounds per square foot per foot of depth when freshly fallen; when wet, its weight will vary with the water content. A commonly used figure is 10 pounds per cubic foot. In calculating snow load, use the depth of the heaviest snow on record multiplied by 10, but never less than 20 pounds per square foot anywhere. If a roof is designed for much less than 20 pounds per square foot of live load, it will not be safe for men to work on.

In the past, there was considerable confusion regarding wind loads against sloping roofs, and some rather weird formulas were evolved to calculate their imaginary intensities, although no one had ever seen a roof blow *in*, and many persons have seen roofs blow *off*. In other words, wind loads are *negative;* that is, they cre-

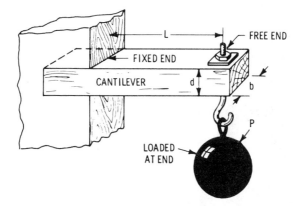

Fig. 4-14. A simple cantilever beam which is fixed at one end and free at the other; this beam is supporting a concentrated load at its free end.

ate *suction*. This is now well recognized, and insurance companies are well informed on the subject.

In regions where hurricanes are not common, the insurance companies recommend that an uplift allowance of 30 pounds per square foot be provided for, acting at right angles to the slope of the roof. This will include the slight *positive* pressure commonly found inside buildings during windstorms. If eaves are wide and overhanging, it is recommended that a gross lifting force of 45 pounds

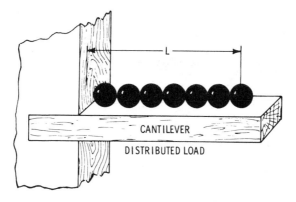

Fig. 4-15. A cantilever beam with an equally distributed load.

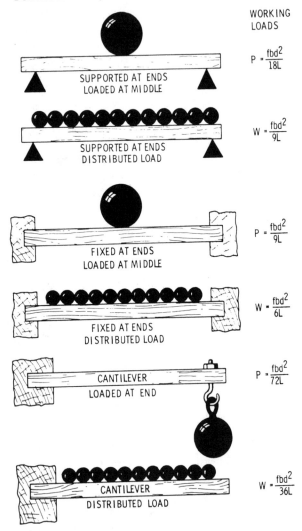

WORKING
LOADS

$P = \dfrac{fbd^2}{18L}$

SUPPORTED AT ENDS
LOADED AT MIDDLE

$W = \dfrac{fbd^2}{9L}$

SUPPORTED AT ENDS
DISTRIBUTED LOAD

$P = \dfrac{fbd^2}{9L}$

FIXED AT ENDS
LOADED AT MIDDLE

$W = \dfrac{fbd^2}{6L}$

FIXED AT ENDS
DISTRIBUTED LOAD

$P = \dfrac{fbd^2}{72L}$

CANTILEVER
LOADED AT END

$W = \dfrac{fbd^2}{36L}$

CANTILEVER
DISTRIBUTED LOAD

Fig. 4-16. The working loads for beams with different types of loads and different modes of support.

per square foot be provided for. If the weight of the roof itself is appreciable, it may be deducted from the gross uplift allowance. It should be recognized, however, that a maximum snow load and a maximum wind load can hardly occur at the same time and thus cannot be considered compensating.

Summary

In most shop work and in the field, the fitness of any species of wood for a given purpose depends on various properties. When treating the strength, stiffness, hardness, and other properties of wood, many factors must be considered. Stiffness is that quality possessed by a beam or other timber to resist the action of a bending force.

A beam is a single structural member, usually horizontal, that carries a load over a given space. The allowable loads on freely supported wood beams of any size can be calculated if the timber is stress-rated and the allowable fiber-stress is known. In many designs, the load is distributed uniformly over the length of the beam. Although this type of loading is actually a series of concentrated loads, it is often found in actual practice, and is usually considered as being uniformly distributed.

Review Questions

1. What is meant by brittleness, bending forces, compression, and elastic limit?
2. What is meant by working stresses?
3. Explain shearing stresses.
4. Why is it important in some cases to distribute the load on a beam between the supports?
5. What is a cantilever beam? Explain its purpose.

CHAPTER 5

Mathematics for Carpenters and Builders

An elementary knowledge of mathematics is essential to the carpenter to solve successfully the numerous problems encountered in almost any branch of carpentry. The branches of mathematics of which the carpenter should possess at least an elementary knowledge are:

1. Arithmetic.
2. Geometry.
3. Trigonometry.

Such knowledge will be found very useful, especially in making up estimates, solving steel square problems, etc.

Arithmetic

By definition arithmetic is *the science of numbers and the art of reaching results by their use.* (Fig. 5-1)

	Notation Basis—1—unit Arithmetic Alphabet 0 1 2 3 4 5 6 7 8 9	
Increased		**Diminished**
By tens 1, 10, 100, 1,000, etc. By varying scales 1 oz. 1 lb. 1 cwt. 1 pt. 1 qt. 1 gal. 1 in. 1 ft. 1 yd.		*By tens* 1, .1, .01, .001, etc. By varying scales $1/4$ $6/7$ $1/3$ oz, $11/23$ etc. $1/6$ lb. $9/8$ oz. $3/8$ cwt. etc.
According to the Four Ground Rules		
Addition Subtraction		Multiplication Division
By involution (powers)		By evolution (roots)
Relations Expressed by		
Ratios 2 : 3 5 : 6 8 : 9 etc. Proportion (equality of ratios) 2 : 3 : : 4 : 6 etc.		
Practical Applications		
Percentage, interest, profit and loss, reduction of weights and measures, measuring, etc.		

Fig. 5-1. Scheme of Arithmetic.

Arithmetic Alphabet

In arithmetic figures are used to represent quantities or magnitudes, thus:

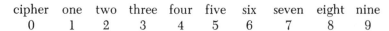

cipher	one	two	three	four	five	six	seven	eight	nine
0	1	2	3	4	5	6	7	8	9

A number is one or a collection of these figures to represent a definite quantity or magnitude as 1 21 517 43,988, etc.

There are various kinds of numbers: simple, compound, integer, abstract, concrete, odd, even, prime, composite, etc., later explained.

Notation and Numeration

By definition, *notation* in arithmetic is *the writing down of figures to express a number*, and numeration is *the reading of the number or collection of figures already written.*

By means of the ten figures given above any number can be expressed.

Figures have two values, simple and local. The simple value of a figure is its value when standing in units' place. The local value of a figure is the value which arises from its location.

When one of the figures stands by itself, it is called a *unit;* but if two of them stand together, the right hand one is still called a unit, but the left hand one is called *tens;* thus, 79 is a collection of 9 units and 7 sets of ten units each, or of 9 units and 70 units, or of 79 units, and is read as seventy-nine.

If three of them stand together, then the left hand one is called *hundreds;* thus 279 is read two hundred and seventy-nine.

To express larger numbers other orders of units are formed, the figure in the 4th place denoting *thousands;* in the 5th place, *ten thousands;* these are called units of the fifth order.

The sixth place denotes hundred thousands, the seventh place denotes millions, etc.

The French method (which is the same as that used in the U.S.) of writing and reading large numbers is shown in the following

TABLE 5-1. Numeration

Names of periods	Billions			Millions			Thousands			Units				Thousandths		
Order of Units	Hundred-billions	Ten-billions	Billions	Hundred-millions	Ten-millions	Millions	Hundred-thousands	Ten-thousands	Thousands	Hundreds	Tens	Units	Decimal point	Tenths	Hundredths	Thousandths
	8	7	6,	5	4	3,	2	0	1,	2	8	2,	•	4	8	9

This system is called Arabic notation, and is the system in ordinary everyday use.

NOTE.—*Roman Notation.* This system is occasionally used as, in the Bible, for chapter headings, corner stones, etc. The method of expressing numbers is by letters, thus:

Roman Table

I Denotes One	XII denotes Twelve	L denotes Fifty
II denotes Two	XIII denotes Thirteen	LX denotes Sixty
III denotes Three	XIV denotes Fourteen	LXX denotes Seventy
IV denotes Four	XV denotes Fifteen	LXXX denotes Eighty
V denotes Five	XVI denotes Sixteen	XC denotes Ninety
VI denotes Six	XVII denotes Seventeen	C denotes One hundred
VII denotes Seven	XVIII denotes Eighteen	D denotes Five hundred
VIII denotes Eight	XIX denotes Nineteen	M denotes One thousand
IX denotes Nine	XX denotes Twenty	$\overline{\text{X}}$ denotes Ten thousand
X denotes Ten	XXX denotes Thirty	$\overline{\text{M}}$ denotes One million
XI denotes Eleven	XL denotes Forty	

In the Roman notation, when any character is placed at the right hand of a larger numeral, its value is added to that of such numeral; as VI, that is, V + I; XV, that is, X + V; MD, that is, M + D; and the like. I, X, and rarely C, are also placed at the left hand of other and larger numerals, and when so situated their value is subtracted from that of such numerals as, IV, that, V − I; XC, that is, C − X; and the like. Formerly the smaller figure was sometimes repeated in such a position twice, its value being in such cases subtracted from the larger; as, IIX, that is, X − II, XXC, that is, C − XX; and the like. Sometimes after the sign IↃ for D, the character Ↄ was repeated one or more times, each repetition having the effect to multiply IↃ by ten; as, IↃↃ, 5,000; IↃↃↃ, 50,000; and the like. To represent numbers twice as great as these, C was repeated as many times before the stroke I, as the Ↄ was after it; as, CCIↃↃ, 10,000; CCCIↃↃↃ, 100,000; and the like. *The ridiculous custom* of using the Roman notation for chapter numbers, year of copyright, sections, etc., should be discontinued.

Definitions

1. Arithmetic is the art of calculating by using numbers.
2. A *number* is a total, amount, or aggregate of units. By counting the units, we arrive at a certain number, such as *two* horses or *five* dozen.
3. A *unit* may mean a single article, but often it means a definite group adopted as a standard of measurement, such as *dozen*,

ton, foot, bushel, or *mile.* Most commonly used units are standardized, and are defined and fixed by law.

4. A *concrete* number is a number applied to some particular unit, such as *ten* nails, *two* dozen eggs, *six* miles.

5. An *abstract number* is one that is *not* applied to any object or group, such as simply *two, four, ten.*

6. *Notation* is the art of expressing numbers by figures or letters. Our system of notation is the Arabic notation. The Roman notation uses letters, such as $V = 5, X = 10$.

7. *Cardinal numbers* are numbers used in simple counting or in reply to the question "How many?" Any number may be a cardinal number.

8. *Ordinal numbers* indicate succession or order of arrangement, such as *first, second, tenth.*

9. An *integer,* or *integral number,* is a whole number, not a fraction or part.

10. An *even number* is any number that can be exactly divided by 2, such as 4, 16, 96, 102.

11. An *odd number* is any number that is *not* exactly divisible by 2, such as 3, 15, 49, 103.

12. A *factor* of a number is a whole number that may be exactly divided into the number. For example, 3 is a factor of 27, 13 is a factor of 91.

13. A *prime number* is a number that has no factors other than itself and 1. Thus, 3, 5, 7, and 23 are prime numbers.

14. A *composite number* is a number that has factors other than itself and 1, such as 8, 49, and 100.

15. A *multiple* of a number is a number that is exactly divisible by a given number. For example, 91 is a multiple of 7, 12 is a multiple of 3.

16. A *digit* is any number from 1 to 9, and usually 0.

Signs of Operation

1. The *sign of addition* is +, and it is read "plus" or "add." Thus, $7 + 3$ is read "seven plus three." The numbers may be taken in *any order* when adding; $7 + 3$ is the same as $3 + 7$.

2. The *sign of subtraction* is −, and it is read "minus." A series of

subtractions *must* be taken in the order written—11 − 7 is *not* the same as 7 − 11.

3. The *sign of multiplication* is ×, and it is read "times" or "multiplied by." The numbers may be taken in *any order* when multiplying; 4 × 7 is the same as 7 × 4.

4. The sign of division is ÷, and it is read "divided by." A series of divisions *must* be taken in the order written—(100 ÷ 2) ÷ 10 = 5.

5. The *sign of equality* is =, and it is read "equals" or "is equal to." The expressions on each side of an equality sign must be numerically the same. The complete expression is called an *equation*.

Use of the Signs of Operation

Example—The use of the sign of addition. A builder when building a house buys 1762 board feet of lumber from one yard, 2176 board feet from another, and 276 board feet from another. How many board feet did he buy?

The problem: 1762 + 2176 + 276 = ?

The solution:
$$
\begin{array}{r}
1762 \\
2176 \\
+\ 276 \\
\hline
4214 \text{ board feet}
\end{array}
$$

Note how the numbers are aligned to permit addition. The units are all aligned on the right, then the tens, then the hundreds, then the thousands, each in the proper column.

Example—The use of the sign of subtraction. A carpenter bought 300 pounds of nails for a job, and he had 28 pounds left when he finished. How many pounds of nails did he use to complete the job?

The problem: 300 − 28 = ?

The solution:

$$
\begin{array}{r}
300 \\
-\ 28 \\
\hline
272 \text{ pounds}
\end{array}
$$

Note that the numbers must be aligned as they were for addition.

Example—*The use of the sign of division.* A carpenter's pickup truck gets an average of 17 miles per gallon of gasoline. How many gallons of gasoline would be required for him to travel 2040 miles in his truck?

The problem: $2040 \div 17 = ?$

The solution:

$$
\begin{array}{r}
120 \\
17\,\overline{)\,2040} \\
\underline{17} \\
34 \\
\underline{34} \\
00
\end{array}
$$

Example—*The use of the sign of equality.* A road contractor finds that he can lay the same amount of paving in 12 days using a 6-man crew that he can lay in 8 days using a 9-man crew. Express this statement as an equation.

$$12 \times 6 = 9 \times 8, \text{ or } 72 = 72$$

Fractions

A fraction indicates that a number or a unit has been divided into a certain number of equal parts, and shows how many of these parts are to be considered. Two forms of fractions are in common usage—the decimal, which is expressed in the tenths system, and the common fraction. The common fraction is written by using two numbers, one written over or alongside the other with a line between them. The lower (or second) number, called the *denomi-*

nator, indicates the number of parts into which the unit has been divided, and the upper (or first) number, called the *numerator*, indicates the number of parts to be considered. In the fraction ⅔, the denominator shows that the unit is divided into 3 parts; the numerator indicates that 2 parts are being considered (Fig. 5–2).

If the quantity indicated by the fraction is less than 1, such as ½, ¾, or ⅚, it is called a *proper fraction*. If the quantity indicated by the fraction is equal to or greater than 1, such as ⅔, ⁵⁄₄, or ⁷⁄₆, it is called an *improper fraction*. When a whole number and a proper fraction are combined, such as 2¼, 6½, it is called a *mixed number*.

Addition of Fractions—Fractions cannot be added without first reducing them to a common denominator.

Example—Add ¾ + ²⁄₉ + ⅔ + ⁷⁄₁₂

To find the common denominator, place the denominators in a row, separated by dashes. Divide them by a prime number that

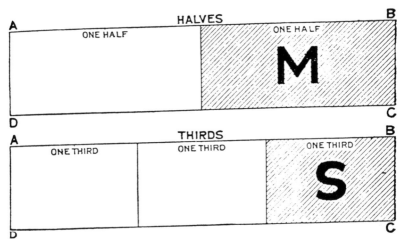

Fig. 5–2. Graphic representation of fractional parts. The figures show a rectangle ABCD, representing a unit divided into two equal parts or halves and into three equal parts or thirds. Evidently the shaded section M, or one half is larger than the shaded section S, or one third.

will divide into at least two of them without leaving a remainder, and bring down the denominators with the dividends that did not contain the divisor without a remainder. Repeat this process as often as possible until there are no two numbers remaining that can be divided by the same number. Then, multiply the divisors and the remainders together, and the result will be the smallest common denominator.

$$
\begin{array}{rl}
\text{The solution:} & 2 \quad) \ \underline{4\text{--}9\text{--}3\text{--}12} \\
& 3 \quad) \ \underline{2\text{--}9\text{--}3\text{--}6} \\
& 2 \quad) \ \underline{2\text{--}3\text{--}1\text{--}2} \\
& 1\text{--}3\text{--}1\text{--}1
\end{array}
$$

The common denominator will then be $2 \times 3 \times 2 \times 3 = 36$. The fractions are then reduced to the common denominator of 36 by multiplying the numerator and the denominator by the same number that will produce 36 in the denominator, as:

$$
\begin{aligned}
\tfrac{3}{4} &= \tfrac{3}{4} \times \tfrac{9}{9} = \tfrac{27}{36} \\
\tfrac{2}{9} &= \tfrac{2}{9} \times \tfrac{4}{4} = \tfrac{8}{36} \\
\tfrac{2}{3} &= \tfrac{2}{3} \times \tfrac{12}{12} = \tfrac{24}{36} \\
\tfrac{7}{12} &= \tfrac{7}{12} \times \tfrac{3}{3} = \tfrac{21}{36}
\end{aligned}
$$

The sum of the fractions $= \dfrac{27 + 8 + 24 + 21}{36} = \tfrac{80}{36} \text{ or } \tfrac{20}{9}$

Multiplication of Common Fractions—A fraction may be multiplied by a whole number by multiplying the numerator of the fraction by that number.

Example—If $\tfrac{3}{4}$ of a keg of nails is used for siding a garage, how many kegs of nails will be used when siding 8 similar garages?

The solution: $8 \times \tfrac{3}{4} = \tfrac{24}{4}$

A fraction may be simplified by dividing both the numerator and the denominator by the same number, and its value will not be affected.

Example—Divide both the numerator and the denominator of the improper fraction $^{24}/_4$ by 4.

The result will be $^6/_1$, or 6. Therefore, 6 kegs of nails will be required to put the siding on the 8 garages.

Fractions may be multiplied by fractions by multiplying their numerators together and their denominators together.

Example—Multiply

$$^2/_5 \times ^1/_4 \times ^5/_{12} = \frac{2 \times 1 \times 5}{5 \times 4 \times 12} = \frac{10}{240} = \frac{1}{24}$$

Multiplication of Fractions by Cancellation—This may readily be done because any factor below the line may be divided by any factor above the line, and any factor above the line may be divided by any factor below the line, without altering the overall value of the expression. Take the example $^{12}/_{30} \times ^{14}/_{56} \times ^{10}/_{24}$ and write it in this form:

$$\frac{1 \times 1 \times 1}{\cancel{12} \times \cancel{14} \times \cancel{10}}$$
$$\frac{\cancel{30} \times \cancel{56} \times \cancel{24}}{3 \times 4 \times 2}$$

The 30 below the line may be divided by the 10 above the line—result, 3. The 56 below the line may be divided by the 14 above the line—result, 4. The 24 below the line may be divided by the 12 above the line—result, 2. The result of the cancellation, then, is

$$\frac{1}{3 \times 4 \times 2} = \frac{1}{24}$$

Division of Fractions—Fractions may be divided by whole numbers by dividing the numerator by that number or by multiplying the denominator by that number.

Example—

$$\frac{7}{8} \div 7 = \frac{1}{8}$$

$$\frac{7}{8} \div 7 = \frac{7}{56} = \frac{1}{8}$$

Fractions may be divided by fractions by inverting the divisor and multiplying.

Example—

$$\frac{7}{8} \div \frac{2}{7} = \frac{7}{8} \times \frac{7}{2} = \frac{49}{16}$$

which is the mixed number $3\frac{1}{16}$.

Subtraction of Fractions—Fractions cannot be subtracted from fractions without first reducing them to a common denominator, as is done for the addition of fractions.

Example—Subtract $\frac{13}{16}$ from $\frac{5}{6}$

Finding the least common denominator,

$$\begin{array}{r} 2\,)\,\overline{16 - 6} \\ \hline 8 - 3 \end{array} \qquad 2 \times 8 \times 3 = 48$$

$$\frac{13}{16} = \frac{13}{16} \times \frac{3}{3} = \frac{39}{48}$$

$$\frac{5}{6} = \frac{5}{6} \times \frac{8}{8} = \frac{40}{48}$$

$$\frac{40 - 39}{48} = \frac{1}{48}$$

To subtract a mixed number from another mixed number, it is usually most convenient to reduce both numbers to improper fractions and then proceed as shown in the last example.

To subtract a mixed number from a whole number, borrow 1

from the minuend, or upper number; reduce the 1 to an improper denominator of the fraction in the subtrahend, or lower number, thereby reducing the whole number by 1. Then make the subtraction in the normal manner.

Example—Subtract $6\frac{7}{8}$ from 14.

The solution: $14 - 6\frac{7}{8} = 13\frac{8}{8} - 6\frac{7}{8} = 7\frac{1}{8}$

Applications of Cancellation—There are countless applications where this method will save appreciable time and work, but care and thought must be given to the proper arrangement of the fractional expression if there are many factors. Also, it must be remembered that if addition or subtraction signs appear, cancellation *may not* be used.

Example—A circular saw has 75 teeth with a 1-inch spacing between each tooth. In order to do satisfactory work, the rim of the saw should travel at approximately 9000 feet per minute. How many revolutions should this saw make per minute? (Hint: 75 inches = $\frac{75}{12}$ feet.)

The solution:

$$\frac{\overset{120}{\cancel{9000}} \times 12}{\cancel{75}} = 120 \times 12 = 1440 \text{ rpm}$$

Example—If you go to a bank and borrow $1000 to purchase a truck, how much will the interest be, at 6% per annum (a theoretical example only), on the money you borrow for 1 year and 3 months? (Hint: 1 year 3 months = 15 months.)

The solution:

$$\frac{\overset{5}{\cancel{6}} \times \overset{\overset{5}{\cancel{10}}}{\cancel{1000}} \times 15}{\underset{2}{\cancel{100} \times \cancel{12}}} = 5 \times 15 = \$75$$

Example—If 8 men in fifteen 8-hour days can throw 1000 cubic yards of gravel into wheelbarrows, how many men will be required to throw 2000 cubic yards of gravel into wheelbarrows in twenty days of 6 hours each?

The solution:

$$\frac{\overset{3}{\cancel{15}} \times \overset{2}{\cancel{8}} \times \overset{4}{\cancel{8}} \times \overset{2}{\cancel{2000}}}{\underset{4}{\cancel{20}} \times \cancel{6} \times \underset{2}{\cancel{1000}}} = 2 \times 4 \times 2 = 16 \text{ men}$$

Example—A building that is 30 feet × 30 feet with a 10-foot ceiling contains approximately 700 pounds of air. What will be the weight of the air in a room 120 feet long, 90 feet wide, and 16 feet high?

The solution:

$$\frac{\overset{4}{\cancel{120}} \times \overset{3}{\cancel{90}} \times 16 \times \overset{70}{\cancel{700}}}{\cancel{30} \times \cancel{30} \times \cancel{10}} = 4 \times 3 \times 16 \times 70 = 13,440 \text{ pounds}$$

Decimals

Decimal means numbering that proceeds by *tens*, and decimal fractions, usually simply called *decimals*, are formed when a unit is divided into 10 parts. When decimals are written, the point where the numbers start is called the *decimal point*. To the left of the decimal point, the numbers read in the regular manner—units, tens, hundreds, thousands, ten thousands, hundred thousands, millions, etc. To the right of the decimal point, the figures are fractional, reading, from the point, tenths, hundredths, thousandths, ten thousandths, hundred thousandths, millionths, etc. (Fig. 5–3).

The common fraction $^6/_{10}$ can be expressed decimally as .6, and the fraction $^{105}/_{1000}$ can be written as .105. The mixed number $106^6/_{100}$ may be expressed decimally as 106.06. The decimal .6 is read "six-tenths," the same as the common fraction $^6/_{10}$, and the dec-

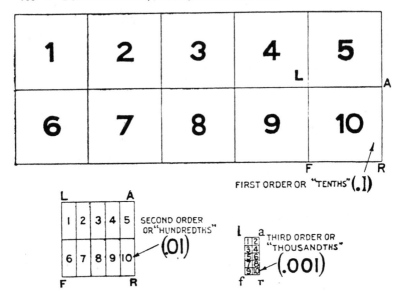

Fig. 5-3. Graphic representation of decimal fractions. A unit divided into ten parts—1st order or "tens"; one of the "tens" as LARF, divided into ten parts—2nd order or "hundredths"; one of the "hundredths" as *larf*, divided into ten parts—3rd order or thousandths. Similarly the process of division may be continued indefinitely.

imal 106.06 is read "one hundred six *and* six-hundredths," the same as the mixed number.

It is not necessary to place a zero before the decimal point, as 0.06, but it is sometimes convenient when it is necessary to align a column of decimals for addition. The decimal points must *always* be aligned for addition, as shown in the following example:

$$
\begin{array}{r}
.6 \\
6.29 \\
+\,10.72 \\
\hline
17.61
\end{array}
$$

The position of the decimal point in the sum is established directly under the column of decimal points above the line. *Note:* the number 327 is *not* read "three hundred *and* twenty-seven," but "three hundred twenty-seven"; however, the decimal 300.27 is read "three hundred *and* twenty-seven hundredths." The decimal .327 is read "three hundred twenty-seven thousandths," while the decimal 300.027 is read "three hundred *and* twenty-seven thousandths."

Reduction of Common Fractions to Decimals—Divide the numerator by the denominator, adding zeros and carrying the division to as many decimal places as are necessary or desirable (Table 5–2).

Example—Reduce the common fraction $^{21}/_{32}$ to a decimal.

$$
\begin{array}{r}
.65625 \\
32\overline{\smash{)}21.00000} \\
\underline{192} \\
180 \\
\underline{160} \\
200 \\
\underline{192} \\
80 \\
\underline{64} \\
160 \\
\underline{160} \\
00
\end{array}
$$

Count off as many decimal places in the quotient as those in the dividend *exceed* those in the divisor. The quotient is .65625.

Subtraction of Decimals—Align the decimal points in the minuend and the subtrahend as shown for addition, and proceed as explained in subtraction of whole numbers. The decimal point in the remainder is placed in exact alignment with those decimal points above the line.

Example—Subtract 27.267 from 167.02.

TABLE 5-2. Fractions of an Inch and Decimal Equivalents

1/2	1/4	1/8	1/16	1/32	1/64	Decimal
					1/64	.015625
				1/32		.03125
					3/64	.046875
			1/16			.0625
					5/64	.078125
				3/32		.09375
					7/64	.109375
		1/8				.125
					9/64	.140625
				5/32		.15625
					11/64	.171875
			3/16			.1875
					13/64	.203125
				7/32		.21875
					15/64	.234875
	1/4					.250
					17/64	.265625
				9/32		.28125
					19/64	.296875
			5/16			.3125
					21/64	.328125
				11/32		.34375
					23/64	.359375
		3/8				.375
					25/64	.390625
				13/32		.40625
					27/64	.421875
			7/16			.4375
					29/64	.453125
				15/32		.46875
					31/64	.484375
1/2						.500
					33/64	.515625
				17/32		.53125
					35/64	.546875
			9/16			.5625
					37/64	.578125
				19/32		.59375
					39/64	.609375
		5/8				.625
					41/64	.640625
				21/32		.65625
					43/64	.671875
			11/16			.6875
					45/64	.703125
				23/32		.71875
					47/64	.734375
	3/4					.750
					49/64	.765625
				25/32		.78125
					51/64	.796875
			13/16			.8125
					53/64	.828125
				27/32		.84375
					55/64	.859375
		7/8				.875
					57/64	.890625
				29/32		.90625
					59/64	.921875
			15/16			.9375
					61/64	.953125
				31/32		.96875
					63/64	.984375
					64/64	1.0000

The solution:

$$167.020$$
$$-\ \ 27.267$$
$$139.753$$

Note that it is necessary to add a zero to the decimal 167.02, in order to make the subtraction, but this does not change its value— $\frac{20}{1000}$ is the same value as $\frac{2}{100}$.

Multiplication of Decimals—Proceed as in multiplication of whole numbers, and count off as many decimal places in the product as there are in *both* the multiplier and multiplicand.

Example—Multiply 1.76 by .06.

The solution:

$$1.76$$
$$\times\ .06$$
$$.1056$$

Compound Numbers

A compound number expresses units of two or more denominations of the same kind, such as 5 yards, 1 foot, and 4 inches. The process of changing the denomination in which a quantity is expressed without changing its value is called reduction. Thus, 1 yard and 2 inches = 38 inches, 25 inches = 2 feet and 1 inch, etc., are examples of reduction. Problems of reduction occur and are explained with the various measures and weights.

Reduction Descending—To reduce a compound number to a lower denomination, multiply the largest units in the given number by the number of units in the next lower denomination, and add to the product the units of that denomination in the given number. Continue this process until the original number is reduced as far as desired. For an explanation of this rule, see the following example.

Example—Reduce the quantity 6 yards, 2 feet, 7 inches to inches.

$$
\begin{array}{rl}
 & 6 \text{ yards} \\
\text{Multiply.} \dots \times\ 3 & \\
 & 18 \text{ feet} \\
\text{Add.} \dots\dots + \ 2 & \\
 & 20 \text{ feet} \\
\text{Multiply.} \dots \times 12 & \\
 & 240 \text{ inches} \\
\text{Add.} \dots\dots + \ 7 & \\
\text{Total.} \dots\dots\dots\ 247 \text{ inches} &
\end{array}
$$

Reduction Ascending—To reduce a number of small units to units of larger denominations, divide the number by the number of units in a unit of the next higher denomination. The quotient is in the higher denomination and the remainder, if any, is in the lower. Continue this process until the number is reduced as far as is desired.

Example—Reduce 378 inches to a quantity of yards, feet, and inches.

The solution:

$$
\begin{array}{l}
12\ )\ 378 \\
\ \ 3\underline{)31}\ \text{feet, 6 inches remainder} \\
\ \ \ 10\ \text{yards, 1 foot remainder}
\end{array}
$$

Therefore, 378 inches = 10 yards, 1 foot, 6 inches.

Ratios

By definition, a ratio is the relation of one number to another as obtained by dividing the first number by the second. Thus, the ratio of 2 to 4 is expressed as 2 : 4; the symbol : is read "to" in the case of a ratio and "is to" in the case of a proportion. It is equivalent to "divided by," hence:

$$2 : 4 = \tfrac{1}{2}$$

The first term of a ratio is the *antecedent*, and the second term is the *consequent*, thus:

$$\begin{array}{ccc} \text{antecedent} & & \text{consequent} \\ 2 & : & 4 \end{array}$$

Since a ratio is essentially a fraction, it follows that if both terms are multiplied or divided by the same number, the value of the ratio is not altered. Thus:

$$2 : 4 = 2 \times 2 : 4 \times 2 = 2 \div 2 : 4 \div 2$$

Two quantities of different kinds cannot form the terms of a ratio. Thus, no ratio can exist between \$5 and 1 day, but a ratio can exist between \$5 and \$2 or between 1 day and 10 days.

Proportion

When two ratios are equal, the four terms form a proportion. A proportion is therefore expressed by using the sign $=$ or $:$: between two ratios, thus:

$$\text{(expressed)} \ 4 : 8 : : 2 : 4$$
$$\text{(read)} \ 4 \text{ is to } 8 \text{ as } 2 \text{ is to } 4$$

The same proportion is also expressed as follows:

$$\tfrac{4}{8} = \tfrac{2}{4}$$

The first and last terms of a proportion are called the *extremes*, and the middle terms are called the *means*, thus:

$$4 : 8 : : 2 : 4$$

The product of the extremes equals the product of the means. Thus, in proportion

$$4 : 8 = 2 : 4$$
$$4 \times 4 = 8 \times 2$$

Since the equation is not altered by dividing both sides by the same number, the value of any term can be obtained as follows:

$$\frac{4 \times 4}{4} = \frac{8 \times 2}{4}$$

$$4 = 2 \times 2 = 4$$

"Rule of Three"

When three terms of a proportion are given, the method of finding the fourth term is called the "rule of three."

Example—If 5 bundles of shingles cost $100, what will 25 bundles cost?

Let X represent the unknown term in the proportion, and, remembering that each ratio must be made up of like quantities,

$$5 \text{ bundles} : 25 \text{ bundles} = 100 \text{ (\$)} : X \text{ (\$)}$$

Multiplying the extremes by the means,

$$5 \times X = 25 \times 100$$

$$X = \frac{25 \times 100}{5} = \$500$$

Percentage

By definition, percentage means the rate per one hundred, or the proportion in one hundred parts. Therefore $\frac{1}{100}$ of a number is called 1 percent; $\frac{2}{100}$, 2 percent, etc. The symbol % is read as percent; thus 1%, 2%, etc. Carefully note the following explanation with respect to the symbol %. The notation 5% means $\frac{5}{100}$, which, when reduced to a decimal (as is necessary when making a calculation), becomes .05, but .05% has a quite different value; .05% means $\frac{.05}{100}$, which, when reduced to a decimal, becomes .0005, that is, $\frac{5}{100}$ of 1%.

If the decimal has more than two places, the figures that follow the hundredths place signify parts of 1 %.

Example—If the list price of shingles is $90 per 1000, what is the net cost for 1000 shingles with a 5 % discount for cash?

Reduce % rate to a decimal.

$$5\% = \frac{5}{100} = .05$$

Multiply decimal by list price.

$$\$90 \times .05 = \$4.50$$

Subtract product obtained from list price.

$$\$90 - \$4.50 = \$85.50$$

Powers of Numbers (Involution)

The word *involution* means the multiplication of a quantity by itself any number of times, and a *power* is the product arising from this multiplication. Involution, then, is the process of raising a number to a given power. The *square* of a number is its second power: the *cube*, its third power, etc. Thus:

$$\text{square of } 2 = 2 \times 2 = 4$$

$$\text{cube of } 2 = 2 \times 2 \times 2 = 8$$

The power to which a number is raised is indicated by a small "superior" figure called an *exponent*. Thus, in Fig. 5-4, the exponent indicates the number of times the number, or "root," has been multiplied by itself.

Fig. 5-4. The "root," "exponent," and "power" of a number.

Roots of Numbers (Evolution)

The word *evolution* means the operation of extracting a root. The root is a factor that is repeated to produce a power. Thus, in the equation $2 \times 2 \times 2 = 8$, 2 is the root from which the power 8 is produced. This number is indicated by the symbol $\sqrt{}$, called the radical sign, which, when placed over a number, means that the root of that number is to be extracted. Thus:

$\sqrt{4}$ *means that the square root of 4 is to be extracted*

The *index* of the root is a small figure that is placed over the radical sign that denotes what root is to be taken. Thus, $^3\sqrt{9}$ indicates the cube root of 9; $^4\sqrt{16}$ indicates the extraction of the fourth root of 16. When there is no index given, the radical sign alone always means the *square root* is to be extracted from the number under the radical sign.

Sometimes the number under the radical sign is to be raised to a power before extracting the root, as:

$$\sqrt[3]{4^3} = \sqrt[3]{4 \times 4 \times 4} = \sqrt[3]{64}$$

Example—Extract the square root of 186,624.

$$
\begin{array}{r}
\sqrt{18'66'24} \qquad 432 \\
16 \\
83 \ \sqrt{266} \\
249 \\
862 \ \sqrt{1724} \\
\underline{1724}
\end{array}
$$

From the decimal point, count off the given number into periods of two places each. Begin with the last period counted off (18). The largest square that can be divided into 18 is 4; put this down in the quotient, and put the square (16) under the 18. Write down the remainder (2), and bring down the next period (66). Multiply 4 (in the quotient) by 2 for the first number of the next divisor; 8 goes into 26 three times. Place 3 after 4 in the quotient and also after 8 in the divisor. Multiply the 83 by 3, placing the product 249 under

266, and subtract, obtaining the remainder 17. Bring down the last period (24), and proceed as before, obtaining 432 as the square root of 186,624.

Extracting the cube root of a number is a more complicated though similar process, as indicated by the following procedure:

1. Separate the number into groups of three figures each, beginning at the decimal point.
2. Find the greatest cube that can be divided into the left-hand group, and write its root for the first figure of the required root.
3. Cube this root, subtract the result from the left-hand group, and annex the next group to the remainder for a dividend.
4. For a partial divisor, take three times the square of the root already found, considered as hundreds, and divide the dividend by it. The quotient (or the quotient diminished) will be (or be close to) the second figure of the root.
5. To this partial divisor, add three times the product of the first figure of the root, considered as tens, by the second figure, and to this add the square of the second figure. This sum will be the complete divisor.
6. Multiply the complete divisor by the second figure of the root, subtract the product from the dividend, and annex the next group to the remainder for a new dividend.
7. Proceed in this manner until all the groups have been annexed. The result will be the cube root required, as shown in the following example.

Example—Extract the cube root of the number 50,653. The solution:

$$\sqrt[3]{50'653.}\ (\ 37$$

$$\begin{array}{r} 27 \\ \hline 23\,653 \\ 23\,653 \end{array}$$

$$\begin{array}{r} 2700 \\ 630 \\ 49 \\ \hline 3379 \end{array}$$

Therefore, the cube root of 50,653 is 37.

Measures

To *measure* is the act or process of determining the extent, quantity, degree, capacity, dimension, volume, etc., of a substance by comparing it with some fixed standard, which is usually fixed by law. A measure may relate to any of these standards. There are many kinds of measures, and practically all of them are standard, but standards vary in different countries. The measures mentioned in this text are all U.S. standards unless designated otherwise. The study of measurements is sometimes called *mensuration*.

Among the many kinds of measures are the following:

1. Linear—measures of length.
2. Square—used to measure areas.
3. Cubic—used to measure volume, or volumetric contents.
4. Weight—many systems of weights are standard.
5. Time—almost standardized all over the world.
6. Circular or angular—the same all over the world.

Linear Measure

Long Measure

12 inches	=	1 foot		
3 feet	=	1 yard	=	36 inches
5½ yards	=	1 rod	=	16½ feet
40 rods	=	1 furlong	=	660 feet
8 furlongs	=	1 mile	=	5280 feet
3 miles	=	1 league (land)		

The furlong is practically never used, except at race tracks and in some athletic events.

Land Surveyor's Measure

7.92 inches	=	1 link		
100 links	=	1 chain	=	66 feet
10 chains	=	1 furlong	=	660 feet
80 chains	=	1 mile	=	5280 feet

The use of the surveyor's chain, or Gunter's chain, was abandoned in the late 1800's and was superseded by the steel tape which is much more accurate; the chain (meaning 66 feet) is still used by the U.S. General Land Office, however, when surveying very old deeds. The standard surveyor's tape is often called, from habit, a chain. It is 100 feet long and is graduated in feet except for the last foot, which is divided into tenths and hundredths of a foot.

Nautical Measure (U.S. Navy)

6 feet = 1 fathom
120 fathoms = 1 cable length
The International Nautical Mile (adopted in 1954) =
  6076.1033 feet
3 nautical miles = 1 marine league

The *knot* is a measure of speed, *not* of length, and is equivalent to 1 nautical mile per hour. A speed of 16 knots is equal to 16 nautical miles per hour.

Square Measure

Square measure is used to measure areas. In most, but not all cases, linear units are used to measure the two dimensions, length and width, and their product is the area in square units. Expressed as an equation,

$$\text{length} \times \text{width} = \text{area}$$

The two dimensions, length and width, must be measured in the same units, but any unit of linear measurement may be used. If inches are multiplied by inches, the result will be in square inches; if feet are multiplied by feet, the result will be in square feet, etc. (Fig. 5–5).

For the small areas commonly found in everyday life, such as table tops or shelves, the unit most commonly used is the square inch. Plywood and lumber are commonly sold by the square foot. Carpets and other floor coverings and materials and ceilings are measured in square yards. The carpenter measures roofing by the

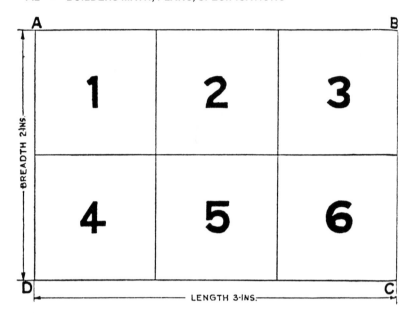

Fig. 5-5. Square measure. If the rectangle ABCD measures 2 inches on one side and 3 inches on the other, and lines are drawn at each inch division, then each of the small squares will have an area of 1 square inch and the area of the rectangle will be area ABCD = breadth × length = 2 × 3 = 6 square inches.

"square" of 10 × 10 feet, or 100 square feet. Tracts of land are usually measured in acres or, for large areas, in square miles.

Square Measure

144 square inches = 1 square foot
9 square feet = 1 square yard
$30^{1}/_{4}$ square yards = 1 square rod = 272.25 square feet
160 square rods = 1 acre = 4840 square yards or 43,560 square feet
640 acres = 1 square mile = 3,097,600 square yards
36 square miles = 1 township

Cubic Measure

Cubic measure is used to determine or appraise volumes. Three dimensions are involved—length, width, and height—and their product is volume. Expressed as an equation,

$$\text{length} \times \text{width} \times \text{height} = \text{volume}$$

As with square measure, the usual linear units—inches, feet, and yards—are ordinarily used to measure these three dimensions. Most small measurements of capacity, such as small shipping cases or small cabinets, are measured in cubic inches. The contents of buildings, their "cubage," are ordinarily expressed in cubic feet. Earthwork, either excavated and loose or in place, is expressed in cubic yards (Fig. 5-6).

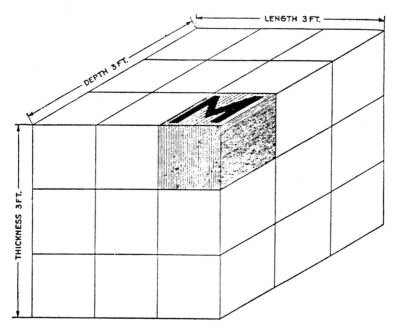

Fig. 5-6.   Cubic measure. If each side of the cube measures 3 feet and it is cut as indicated by the lines, each little cube as M, will have each of its sides 1 foot long and will contain 1 × 1 × 1 = 1 cubic foot. Accordingly the large cube will contain 3 × 3 × 3 = 27 cubic feet or 1 cubic yard.

Cubic Measures of Volume

1728 cubic inches = 1 cubic foot
27 cubic feet = 1 cubic yard

Dry Measure—Quantities of loose, granular materials, such as grains, some fruits, and certain vegetables, are measured in arbitrary units, that, in turn, are defined by means of cubic measures of volume, usually in cubic inches. Their value is sometimes, but not always, fixed by law.

Dry Measure (U.S.)

2 pints = 1 quart = 67.2 cubic inches
8 quarts = 1 peck = 537.61 cubic inches
4 pecks = 1 bushel = 2150.42 cubic inches

Dry Measure (British and Canadian)

1 gallon = .5 peck = 277.42 cubic inches
4 pecks = 1 bushel = 2219.23 cubic inches

The British dry quart is apparently not often used; it is equal to 69.35 cubic inches, or 1.032 U.S. dry quarts.

The weight, rather than the volume, of grains is the standard fixed by the U.S. government:

1 bushel of wheat = 60 pounds
1 bushel of barley = 48 pounds
1 bushel of oats = 32 pounds
1 bushel of rye = 56 pounds
1 bushel of corn (shelled) = 56 pounds

Board or Lumber Measure—Timbers and logs are measured in board or lumber measure. The board foot is 1 foot wide, 1 foot long, and 1 inch thick, thereby containing 144 cubic inches. In the retail market, all lumber which is less than one inch thick is called one inch. At the sawmills, the full sizes govern the thickness of the saw kerfs; usually about ¼ inch is allowed for and accounted as

sawing loss. Actual finished (dressed) sizes of common lumber, the dimension and timbers for pine are as follows:

The standard dressed thickness of 1-inch boards is ¾ inch.

The standard thickness of 2-inch dimension boards is 1½ inches.

The standard dressed widths of lumber 2 inches thick and less are ½ inch less for widths under 8 inches and ¾ inch less for 8-inch widths and wider.

The standard dressed widths and thicknesses for lumber and timbers are ½ inch less both ways under 8 inches wide and ¾ inch for 8-inch widths and over. So a 2″ × 8″ would be 1½″ × 7¼″. A 2″ × 10″ would be 1½″ × 9¼″.

Liquid Measure—Liquid measure is used to measure various liquids such as oils, liquors, molasses, and water.

Liquid Measure

4 gills = 1 pint = 28.875 cubic inches
2 pints = 1 quart = 57.75 cubic inches
4 quarts = 1 gallon = 231 cubic inches

There is no legal standard barrel in the U.S. By custom, a barrel of water is understood to be 31½ gallons. The British barrel is generally 36 Imperial gallons. Crude oil is often disposed of at the wells in barrels of 50 gallons, while refined oils are marketed in barrels of 48 gallons. Owing to this lack of uniformity, it is safest to specify "barrels of 50 gallons," or something of that nature, to avoid misunderstanding. The barrel is sometimes used as a dry measure unit of varying value. For Portland cement, 4 bags = 1 barrel = 4 cubic feet = 376 pounds.

Measures of Weight

The simplest definition of weight is that it is the force with which a body is attracted toward the earth. It is a quantity of heaviness. There are three systems, or standards, of weights used in the United States. They are:

1. Avoirdupois—used for almost all ordinary purposes
2. Troy—used in weighing precious metals and jewels
3. Apothecaries—used by pharmacists when compounding drugs

Avoirdupois Weights

16 drams = 1 ounce
16 ounces = 1 pound
100 pounds = 1 hundredweight
20 hundredweights = 1 ton

In England, the following weights are in common usage:

14 pounds = 1 stone
112 pounds = 1 hundredweight
20 hundredweight = 1 ton = 2240 pounds

The 2240-pound ton is sometimes used in the United States for weighing coal at the mines and at customs houses for evaluating shipments from England.

Troy Weights

3.086 grains = 1 carat
24 grains = 1 pennyweight
20 pennyweights = 1 ounce
12 ounces = 1 pound

Apothecaries Weights

20 grains = 1 scruple
3 scruples = 1 dram
8 drams = 1 ounce
12 ounces = 1 pound

This standard of weights is fast becoming obsolete, although pharmacists must be familiar with it. Manufacturing pharmacists and chemists are rapidly changing to the metric weights, using the metric *gram* as a basis instead of the apothecaries' scruple; 1 scruple = 1.296 grams.

Time Measure

Time is defined as measurable duration. It is the period during which an action or process continues. The basis, or standard, used in our ordinary determination of time is the *mean solar day*, beginning and ending at mean midnight. The word "mean" as used here simply means average; the direct ray of the sun does not move in an exact and uniform path around the equator.

Time Measure

60 seconds	=	1 minute
60 minutes	=	1 hour
24 hours	=	1 day
7 days	=	1 week
30 days (commonly)	=	1 month
365 days	=	1 year
10 years	=	1 decade
100 years	=	1 century
1000 years	=	1 millennium

The length of an *astronomical* year is 365 days, 5 hours, 48 minutes, and 45.51 seconds, or approximately 365¼ days. This makes it necessary to add 1 day every 4 years, thus making the "leap" year 366 days.

Circular Measure

This measure is used in astronomy, land surveying, navigation, and in measuring angles of all kinds. Circles of all sizes are divisible into degrees, minutes, and seconds. Note that a degree is *not* a measurement of length. It is $\frac{1}{360}$ of the circumference of a circle with any radius.

Circular Measure

60 seconds	=	1 minute
60 minutes	=	1 degree
360 degrees	=	1 circle

The Metric System

The base, or fundamental, unit in the metric system is the meter. The meter is defined as the distance between two scribed marks on a standard bar made of platinum-iridium kept in the vaults of the International Bureau of Weights and Measures, near Paris, France. Of course, many other standard meter bars have been made from the measurement on this bar. It is permissible and official to use this measurement in the United States, and, in fact, the yard, the basis for the English system of measurement, has been defined as exactly $\frac{3600}{3937}$ meter, or 1 meter = 39.37 inches.

The advantage, and immeasurably greater convenience, of the metric system over the English system of units lies in the fact that it is expressed in tenths, thereby readily allowing the use of decimals. However, the American public is accustomed to the English units, and as recent experience indicates, the system should continue for a long time. The metric system is, of course, in common use all over the world with the exception of some English-speaking peoples. The meter is used like the yard to measure cloth and short distances.

Units of other denominations are named by prefixing to the word "meter" the Latin numerals for the lower denominations and the Greek numerals for the higher denominations, as shown in the following chart:

Lower denomination			*Higher denomination*		
Deci	=	$\frac{1}{10}$	Deka	=	10
Centi	=	$\frac{1}{100}$	Hecto	=	100
Milli	=	$\frac{1}{1000}$	Kilo	=	1000
Micro	=	$\frac{1}{1,000,000}$	Myria	=	10,000
			Mega	=	1,000,000

Therefore, 1 decimeter = $\frac{1}{10}$ of a meter, 1 millimeter = $\frac{1}{1000}$ of a meter, 1 kilometer = 1000 meters, etc. From this explanation of the metric prefixes, the linear table that follows can easily be understood.

Metric Table of Linear Measure

Metric Denomination			Meter	U.S. value
		1 millimeter =	.001 =	.0394 inches
10 millimeters	=	1 centimeter =	.01 =	.3937 inches
10 centimeters	=	1 decimeter =	.1 =	3.937 inches
10 decimeters	=	1 meter =	1. =	39.3707 inches
				= 3.28 feet
10 meters	=	1 dekameter =	10. =	32.809 feet
10 dekameters	=	1 hectometer =	100. =	328.09 feet
10 hectometers	=	1 kilometer =	1000. =	.62138 miles
10 kilometers	=	1 myriameter =	10,000. =	6.2138 miles

The kilometer is commonly used for measuring long distances. The square meter is the unit used for measuring ordinary surfaces, such as flooring or ceilings.

Metric Table of Square Measure

100 square millimeters (mm²)	= 1 square centimeter	= .15 + square inch
100 square centimeters (cm²)	= 1 square decimeter	= 15.5 + square inches
100 square decimeters (dm²)	= 1 square meter (m²)	= 1.196 + square yards

The acre is the unit of land measure and is defined as a square whose side is 10 meters, equal to a square dekameter, or 119.6 square yards.

Metric Table of Land Measure

1 centiare (ca)	= 1 square meter	= 1.196 square yards
100 centiares (ca.)	= 1 acre	= 119.6 square yards
100 ares (A.)	= 1 hectare	= 2.471 acres
100 hectares (ha.)	= 1 square kilometer	= .3861 square miles

The cubic meter is the unit used for measuring ordinary solids, such as excavations or embankments.

Metric Table of Cubic Measure

1000 cubic millimeters (mm³)	= 1 cubic centimeter	= .061 + cubic inches
1000 cubic centimeters (cm³)	= 1 cubic decimeter	= 61.026 + cubic inches
1000 cubic decimeters (dm³)	= 1 cubic meter	= 35.316 + cubic feet

The liter is the unit of capacity, both of liquid and of dry measures, and is equivalent to a vessel whose volume is equal to a cube whose edge is one-tenth of a meter, equal to 1.0567 quarts liquid measure, and .9081 quart dry measure.

Metric Table of Capacity

10 milliliters (ml.)	= 1 centiliter	= .0338 fluid ounce
10 centiliters (cl.)	= 1 deciliter	= .1025 cubic inch
10 deciliters (dl.)	= 1 liter	= 1.0567 liquid quart
10 liters (l.)	= 1 dekaliter	= 2.64 gallons
10 dekaliters (dl.)	= 1 hectoliter	= 26.418 gallons
10 hectoliters (hl.)	= 1 kiloliter	= 264.18 gallons
10 kiloliters (kl.)	= 1 myrialiter (ml.)	

1 myrialiter	=	10 cubic meters	
		= 283.72 + bushels	= 2641.7 + gallons
1 kiloliter	=	1 cubic meter	
		= 28.372 + bushels	= 264.17 gallons
1 hectoliter	=	1/10 cubic meter	
		= 2.8372 + bushels	= 26.417 gallons
1 decaliter	=	10 cubic decimeters	
		= 9.08 quarts	= 2.6417 gallons
1 liter	=	1 cubic decimeter	
		= .908 quart	= 1.0567 quart liquid
1 deciliter	=	1/10 cubic decimeter	
		= 6.1022 cubic inches	= .845 gallons

1 milliliter	=	10 cubic centimeters		
	=	.6102 cubic inches	=	.338 fluid ounces
1 centiliter	=	1 cubic centimeter		
	=	.061 cubic inches	=	.27 fluid dram

The hectoliter is the unit used for measuring liquids, grain, fruit, and roots in large quantities. The gram is the unit of weight equal to the weight of a cube of distilled water, the edge of which is $\frac{1}{100}$ of a meter and is equal to 15.432 troy grains.

Metric Table of Weight Measure

10 milligrams (mg.)	= 1 centigram	= .15432 + grains troy
10 centigrams (cg).	= 1 decigram	= 1.54324 + grains troy
10 decigrams (dg.)	= 1 gram	= 15.43248 + grains troy
10 grams (g.)	= 1 dekagram	= .35273 + ounce avoirdupois
10 dekagrams (Dg.)	= 1 hectogram	= 3.52739 + ounces avoirdupois
10 hectograms (hg.)	= 1 kilogram	= 2.20462 + pounds avoirdupois
10 kilograms (kg.)	= 1 myriagram	= 22.04621 + pounds avoirdupois
10 myriagrams (Mg.)	= 1 quintal	= 220.46212 + pounds avoirdupois
10 quintals	= 1 ton	= 2204.62125 + pounds avoirdupois

Geometry

By definition, geometry is that branch of mathematics that deals with space and figures in space. In other words, it is the science of the mutual relations of points, lines, angles, surfaces, and solids, which are considered as having no properties except those arising from extension and difference of situation.

Lines

There are two kinds of lines—straight and curved. A straight line is the shortest distance between two points. A curved line is one which changes its direction at every point. Two lines are said to be parallel when they have the same direction. A horizontal line is one parallel to the horizon or surface of the earth. A line is perpendicular with another line when they are at right angles to each other. These definitions are illustrated in Fig. 5–7.

Angles

An angle is the difference in direction between two lines proceeding from the same point (called the vertex). Angles are said to be right (90 degrees) when formed by two perpendicular lines, Fig. 5–8A, acute (less than 90 degrees) when less than a right angle, Fig. 5–8B, and obtuse (more than 90 degrees) when greater than a right

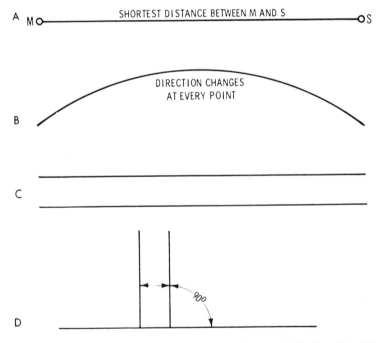

Fig. 5–7. Various lines, (A) Straight; (B) Curved; (C) Parallel; (D) Perpendicular.

angle, Fig. 5–8C. All angles except right (or 90-degree) angles are called oblique angles.

Angles are usually measured in degrees (circular measure) (Fig. 5–8D). The *complement* of an angle is the difference between 90 degrees and the angle; the *supplement* of the angle is the difference between the angle and 180 degrees.

Plane Figures

The term *plane figures* means a plane surface bounded by straight or curved lines, and a plane, or plane surface, is one in which any straight line joining any two points lies wholly in the surface. Fig. 5–9 defines a plane surface. There is a great variety of plane figures, which are known as polygons when their sides are straight lines. The sum of the sides is called the *perimeter*. A regu-

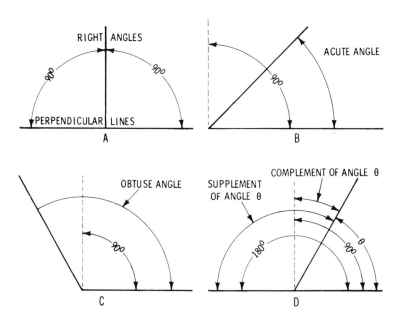

Fig. 5–8. Various angles, (A) Right; (B) Acute; (C) Obtuse; (D) Complement and supplement of an angle.

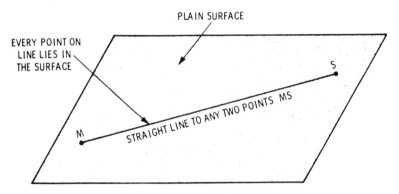

Fig. 5-9.   A plane surface means that every point on a straight line joining any two points in the surface lies in the surface.

lar polygon has all its sides and angles equal. Plane figures of three sides are known as triangles (Fig. 5–10), and plane figures of four sides are quadrilaterals. Fig. 5–11 shows examples of these; Fig. 5–13 details the structure of the quadrilateral. Various plane figures are formed by curved sides and are known as circles, ellipses, etc., as shown in Fig. 5–12.

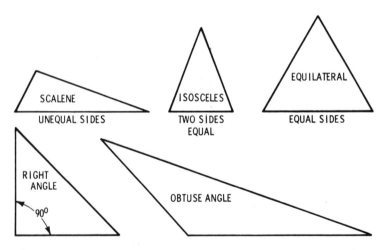

Fig. 5-10.   Various triangles; a triangle is a polygon having three sides and three angles.

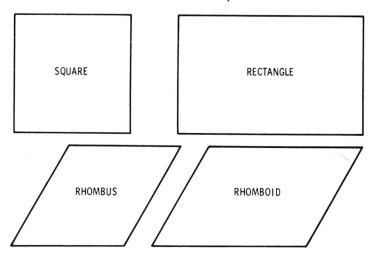

Fig. 5-11. Various quadrilaterals; all opposite sides of a quadrilateral are equal.

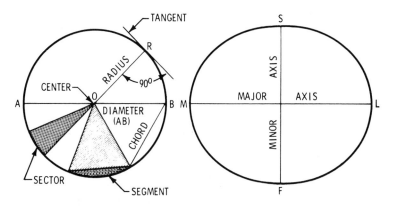

Fig. 5-12. Curved figures. A circle is a plane figure bounded by a uniformly curved line, every point of which is equidistant from the center point O; OR is a radius, and AB is a diameter. An ellipse is a curved figure enclosed by a curved line which is such that the sum of the distances between any point on the circumference and the two foci is invariable; ML is the major axis, and SF is the minor axis.

Solids

Solids have three dimensions—length, width, and thickness. The bounding planes are called the *faces*, and the intersections are called the *edges*. A prism (Fig. 5–14) is a solid whose ends consist of equal and parallel polygons and whose sides are parallelograms. The altitude of a prism is the perpendicular distance of its opposite sides or bases. A parallelopipedon is a prism which is bounded by six parallelograms; the opposite parallelograms are parallel and equal. A cube is a parallelopipedon whose faces are equal. One important solid is the cylinder, which is a body bounded by a uniformly curved surface and having its ends equal and forming par-

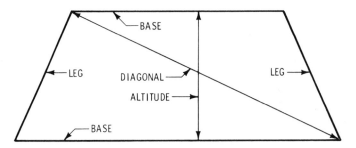

Fig. 5-13. The parallel sides of a quadrilateral (four-sided polygon) are the bases; the distance between the bases is the altitude, and a line joining two opposite vertices is a diagonal.

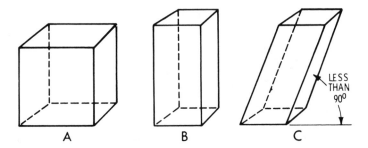

Fig. 5-14. Various prisms, (A) Cube, or equilateral parallelopipedon; (B) Parallelopipedon; (C) Oblique parallelopipedon.

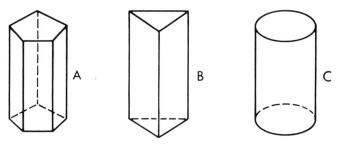

Fig. 5-15. Various solids, (A) Pentagonal prism; (B) Triangular prism; (C) Cylinder.

allel circles (Fig. 5-15). There are numerous other solids having curved surfaces, such as cones and spheres.

Geometrical Problems

The following problems illustrate the method in which various geometrical figures are constructed, and they should be solved by the use of pencil, dividers, compass, and scale. Many of these problems are commonly encountered in carpentry with layout work; therefore, experience in working them out will be of value to carpenters and woodworkers.

Problem 1—To bisect, or divide into two equal parts, a straight line or arc of a circle.

In Fig. 5-16, from the ends A and B, as centers, describe arcs cutting each other at C and D, and draw line CD, which cuts the line at E, or the arc at F.

Problem 2—To draw a perpendicular to a straight line, or a radial line to an arc.

The line CD is perpendicular to AB; also, the line CD is radial to the arc AB (Fig. 5-16).

Problem 3—To erect a perpendicular to a straight line from a given point in that line.

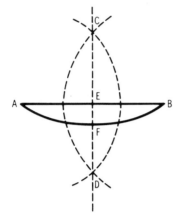

Fig. 5-16. To bisect a straight line or arc of a circle.

In Fig. 5-17, with any radius from any given point A, in the line BC describe arcs cutting the line at B and C. Next, with a longer radius describe arcs with B and C as centers, intersecting at D, and draw the perpendicular DA.

Second Method—In Fig. 5-18, from any point F above BC, describe a circle passing through the given point A and cutting the given line at D; draw DF, and extend it to cut the circle at E; draw the perpendicular AE.

Third Method (boat builders' layout method)—In Fig. 5-19, let MS be the given line and A, the given point. From A, measure

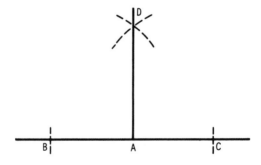

Fig. 5-17. To erect a perpendicular to a straight line from a given point on that line.

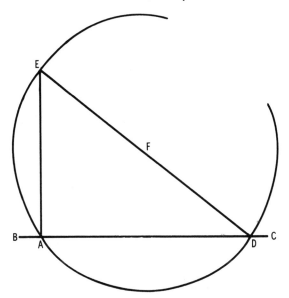

Fig. 5-18. To erect a perpendicular to a straight line from a given point on that line, second method.

off a distance *AB* (4 feet). With centers *A* and *B* and radii of 3 and 5 feet, respectively, describe arcs *L* and *F* intersecting at *C*. Draw a line through *A* and *C*, which will be the perpendicular required. This method is used extensively by carpenters when squaring the corners of buildings, but they ordinarily use multiples of 3, 4, and 5, such as 6, 8, and 10, or 12, 16, and 20.

Fourth Method—In Fig. 5-20, from *A*, describe an arc *EC*, and from *E* with the same radius describe the arc *AC*, cutting the other at *C*; through *C*, draw a line *ECD*; lay off *CD* equal to *CE*, and through *D*, draw the perpendicular *AD*. The triangle produced is exactly 60 degrees at *E*, 30 degrees at *D*, and 90 degrees at *A*. The hypotenuse *ED* is exactly twice the length of the base *EA*.

Problem 4—To erect a perpendicular to a straight line from any point outside the line.

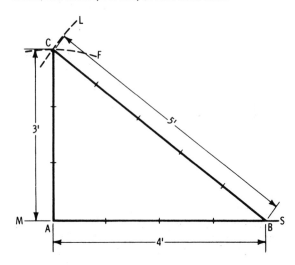

Fig. 5-19. To erect a perpendicular to a straight line from a given point on that line, third method.

In Fig. 5-21, from the point A, with a sufficient radius cut the given line at F and G, and from these points describe arcs cutting at E. Draw the perpendicular AE.

Second Method—In Fig. 5-22, from any two points B and C at some distance apart in the given line and with the radii BA and CA,

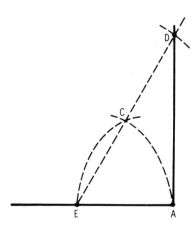

Fig. 5-20. To erect a perpendicular to a straight line from a given point on that line, fourth method.

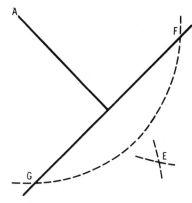

Fig. 5-21. To erect a perpendicular to a straight line from any point outside the line.

respectively, describe arcs cutting at A and D. Draw the perpendicular AD.

Problem 5—To draw a line parallel to a given line through a given point.

In Fig. 5-23, with C as the center, describe an arc tangent to the given line AB; the radius will then equal the distance from the

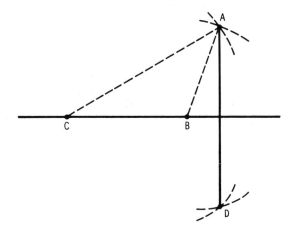

Fig. 5-22. To erect a perpendicular to a straight line from any point outside the line, second method.

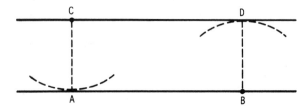

Fig. 5–23.   To draw a line parallel to a given line through a given point.

given point to the given line. Take a point B on the given line remote from C, and describe an arc. Draw a line through C, tangent to this arc at D, and it will be parallel to the given line AB.

Second Method—In Fig. 5–24, from A, the given point, describe the arc FD, cutting the given line at F; from F, with the same radius, describe the arc EA, and lay off FD equal to EA. Draw the parallel line through the points AD.

Problem 6—To divide a line into a number of equal parts.

In Fig. 5–25, assuming line AB is to be divided into five equal parts, draw a diagonal line AC of five units in length. Join BC at 5 and through the points $1, 2, 3, 4$, draw lines $1L, 2a$, etc., parallel to BC. AC will then be divided into five equal parts, AL, La, ar, rf, and fB.

Problem 7—To draw an angle equal to a given angle on a straight line. In Fig. 5–26, let A be the given angle, and FG the line. With any radius from the points A and F, describe arcs DE and IH cutting the sides of angle A and

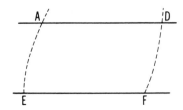

Fig. 5–24.   To draw a line parallel to a given line through a given point, second method.

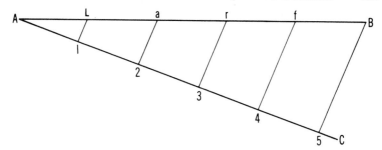

Fig. 5-25. To divide a line into a number of equal parts.

line *FG*. Lay off arc *IH* equal to arc *DE*, and draw line *FH*. Angle *F* is then equal to *A*, as required.

Problem 8—To bisect an angle.

In Fig. 5-27, let *ACB* be the angle; with the center of the angle at *C*, describe an arc cutting the sides at *A* and *B*. Using *A* and *B* as centers, describe arcs which intersect at *D*. A line through *C* and *D* will divide the angle into two equal parts.

Problem 9—To find the center of a circle.

In Fig. 5-28, draw any chord *MS*. With *M* and *S* as centers, and with any radius, describe arcs *L F* and *L′ F′*, and draw a line through their intersection, giving a diameter *AB*. Applying the same construction with centers *A* and *B*, describe arcs *ef* and *e′f′*. A

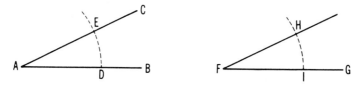

Fig. 5-26. To draw an angle equal to a given angle on a straight line.

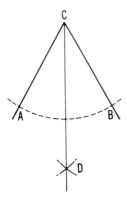

Fig. 5-27. To bisect an angle.

line drawn through the intersections of these arcs will cut line AB at O, the center of the circle.

Problem 10—To describe an arc of a circle with a given radius through two given points.

In Fig. 5-29, take the given points A and B as centers, and, with the given radius, describe arcs which intersect at C; from C, with the same radius, describe an arc AB, as required.

Second Method—In Fig. 5-30, for a circle or an arc, select

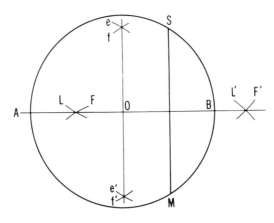

Fig. 5-28. To find the center of a circle.

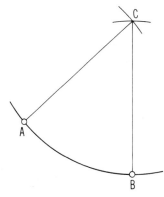

Fig. 5-29. To describe an arc of a circle with a given radius through two given points.

three points *ABC* in the circumference which are well apart; with the same given radius, describe arcs from these three points that intersect each other, and draw two lines, *DE* and *FG*, through their intersections. The point where these lines intersect is the center of the circle or arc.

Problem 11—To describe a circle passing through three given points.

In Fig. 5–30, let *A*, *B*, and *C* be the given points, and proceed as in the last problem to find the center *O* from which the circle may be described. This problem is useful in such work as laying out an

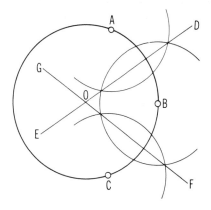

Fig. 5-30. To describe an arc of a circle with a given radius through two given points, second method.

object of large diameter, such as an arch, when the span and rise are given.

Problem 12—To draw a tangent to a circle from a given point in the circumference.

In Fig. 5–31, from *A*, lay off equal segments *AB* and *AD*; join line *BD*, and draw line *AE* parallel to *BD* for the tangent.

Problem 13—To draw tangents to a circle from points outside the circle.

In Fig. 5–32, from *A*, and with the radius *AC*, describe an arc *BCD*; from *C*, with a radius equal to the diameter of the circle, intersect the arc at *BD*; join *BC* and *CD*, which intersect the circle at *E* and *F*, and draw the tangents *AE* and *AF*.

Problem 14—To describe a series of circles tangent to two inclined lines and tangent to each other.

In Fig. 5–33, bisect the inclination of the given lines *AB* and *CD* by the line *NO*. From a point *P* in this line, draw the perpendicular *PB* to the line *AB*, and on *P*, describe the circle *BD*, touching the lines and the center line at *E*. From *E*, draw *EF* perpendicular to the center line intersecting *AB* at *F*, and from *F*, describe an arc *EG* intersecting *AB* at *G*. Draw *GH* parallel to *BP*, thus pro-

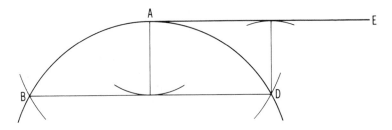

Fig. 5-31. To draw a tangent to a circle from a given point in the circumference.

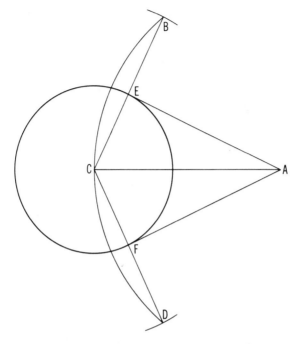

Fig. 5-32. To draw tangents to a circle from points outside the circle.

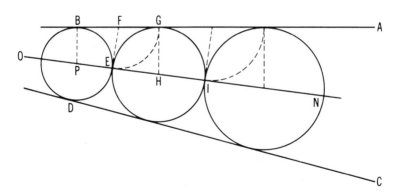

Fig. 5-33. To describe a series of circles tangent to two inclined lines and tangent to each other.

ducing *H*, the center of the next circle, to be described with the radius *HE*, and so on for the next circle *IN*.

Problem 15—To construct an equilateral triangle on a given base.

In Fig. 5–34, with *A* and *B* as centers and a radius equal to *AB*, describe arcs *l* and *f*. At their intersection *C*, draw lines *CA* and *CB*, which are the sides of the required triangle. If the sides are to be unequal, the process is the same, taking as the radii the lengths of the two sides to be drawn.

Problem 16—To construct a square on a given base.

In Fig. 5–35, with end points *A* and *B* of the base as centers and a radius equal to *AB*, describe arcs which intersect at *C*; on *C*, describe arcs which intersect the others at *D* and *E*, and on *D* and *E*, intersect these arcs *F* and *G*. Draw *AE* and *BG*, and join the intersections *HI* to form the square *AHIB*.

Problem 17—To construct a rectangle on a given base.

In Fig. 5–35, let *AB* be the given base. Erect a perpendicular at *A* and at *B* equal to the altitude of the rectangle, and join their ends *F* and *G* by line *FG*; *AFGB* is the rectangle required.

Problem 18—To construct a parallelogram given the sides and an
    angle.

In Fig. 5–36, draw side *DE* equal to the given length *A*, and lay

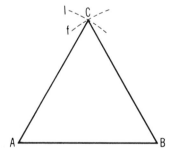

Fig. 5–34. To construct an equilateral triangle on a given base.

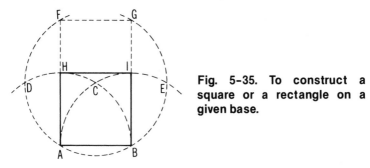

Fig. 5-35. To construct a square or a rectangle on a given base.

off the other side DF, equal to the other length B, thus forming the given angle C. From E, with DF as the radius, describe an arc, and from F, with the radius DE, intersect the arc at G. Draw FG and EG. The remaining sides may also be drawn as parallels to DE and DF.

Problem 19—To draw a circle around a triangle.

In Fig. 5-37, bisect two sides AB and AC of the triangle at E and F, and from these points draw perpendiculars intersecting at K. From K, with radius KA or KC, describe the circle ABC.

Problem 20—To circumscribe and inscribe a circle about a square.

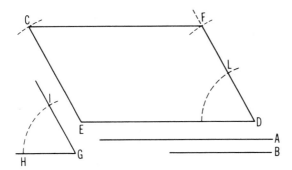

Fig. 5-36. To construct a parallelogram given the sides and an angle.

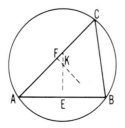

Fig. 5–37. To draw a circle around a triangle.

In Fig. 5–38, draw the diagonals *AB* and *CD* intersecting at *E*. With a radius *EA*, circumscribe the circle. To inscribe a circle, draw a perpendicular from the center (as just found) to one side of the square, as line *OM*. With radius *OM*, inscribe the circle.

Problem 21 — To circumscribe a square around a circle.

In Fig. 5–39, draw diameters *MS* and *LF* at right angles to each other. At points *M*, *L*, *S*, and *F*, where these diameters intersect the circle, draw tangents, that is, lines perpendicular to the diameters, obtaining the sides of the circumscribed square *ABCD*.

Problem 22 — To inscribe a circle in a triangle.

In Fig. 5–40, bisect two angles *A* and *C* of the triangle with

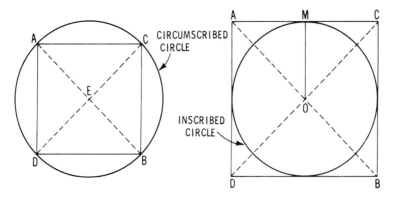

Fig. 5–38. To circumscribe and inscribe a circle around, inside a square.

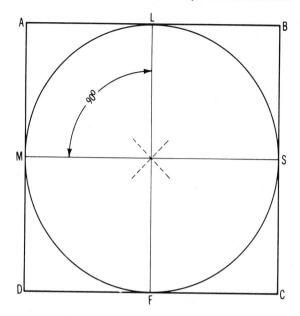

Fig. 5-39. To circumscribe a square about a circle.

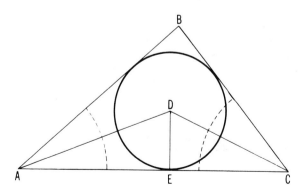

Fig. 5-40. To inscribe a circle in a triangle.

lines that intersect at D; from D, draw a perpendicular DE to any side. With DE as the radius, describe a circle.

Problem 23—To inscribe a pentagon in a circle.

In Fig. 5–41, draw two diameters AC and BD at right angles intersecting at O; bisect AO at E, and from E, with radius EB, AC at F; from B, with radius BF, intersect the circumference at G and H, and with the same radius, step round the circle to I and K; join the points thus found to form the pentagon $BGIKH$.

Problem 23A—To inscribe a five-pointed star in a circle.

In Fig. 5–42, proceed as explained for the inscribed pentagon in Problem 23. Then, connect point B with points K and I, point H with points G and I, etc. The star is mathematically correct.

Problem 24—To construct a hexagon from a given straight line.

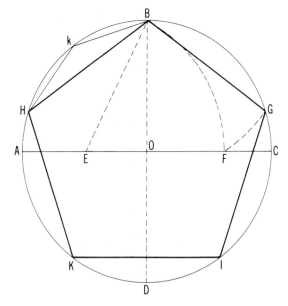

Fig. 5-41.   To inscribe a pentagon in a circle.

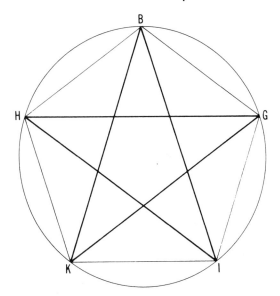

Fig. 5-42. To inscribe a five-pointed star in a circle.

In Fig. 5-43, from *A* and *B*, the ends of the given line, describe arcs intersecting at *g*; from *g*, with the radius *gA*, describe a circle. With the same radius, lay off arcs *AG*, *GF*, *BD*, and *DE*. Join the points thus found to form the hexagon.

Problem 25—To inscribe a hexagon in a circle.

In Fig. 5-44, draw a diameter *ACB*; from *A* and *B*, as centers with the radius of the circle *AC*, intersect the circumference at *D*, *E*, *F*, and *G*, and draw lines *AD*, *DE*, etc., to form the hexagon. The points *D*, *E*, etc., may also be found by stepping off the radius (with the dividers) six times around the circle.

Problem 26—To describe an octagon on a given straight line.

In Fig. 5-45, extend the given line *AB* both ways, and draw perpendiculars *AE* and *BF*; bisect the external angles *A* and *B* by lines *AH* and *BC*, which are made equal to line *AB*. Draw *CD* and

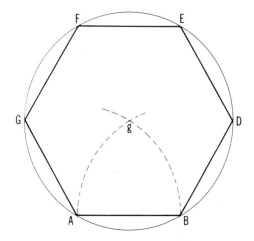

Fig. 5–43. To construct a hexagon from a given straight line.

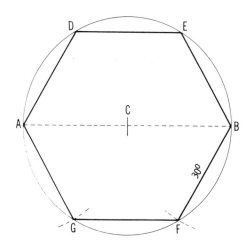

Fig. 5–44. To inscribe a hexagon in a circle.

HG parallel to *AE* and equal to line *AB*. Draw *CD* and *HG* parallel to *AE* and equal to line *AB*; with *G* and *D* as centers, and with the radius equal to *AB*, intersect the perpendiculars at *E* and *F*, and draw line *EF* to complete the hexagon.

Problem 27—To inscribe an octagon in a square.

In Fig. 5–46, draw the diagonals of the square intersecting at *e*; from the corners *A*, *B*, *C*, *D*, with *Ae* as the radius, describe arcs intersecting the sides of the square at *g*, *h*, etc., and join the points found to complete the octagon.

Problem 28—To inscribe an octagon in a circle.

In Fig. 5–47, draw two diameters *AC* and *BD* at right angles; bisect the arcs *AB*, *BC*, etc., at *e*, *f*, etc., to form the octagon.

Problem 29—To circumscribe an octagon about a circle.

In Fig. 5–48, describe a square about the given circle *AB*; draw perpendiculars *hk*, etc., to the diagonals, touching the circle, to

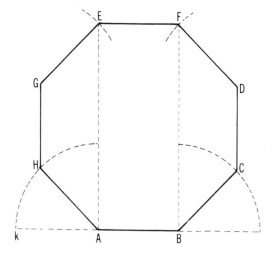

Fig. 5–45.   To describe an octagon on a given straight line.

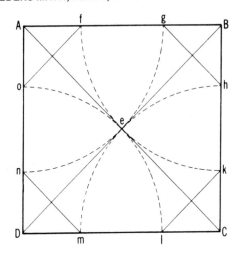

Fig. 5-46.   To inscribe an octagon in a square.

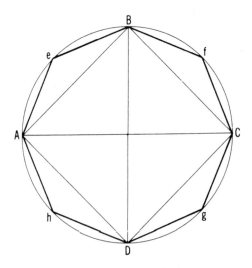

Fig. 5-47.   To inscribe an octagon in a circle.

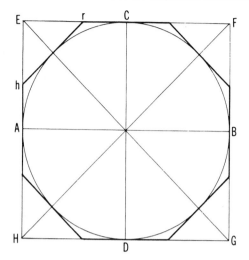

Fig. 5-48. To circumscribe an octagon about a circle.

form the octagon. The points h, k, etc., may be found by cutting the sides from the corners.

Problem 30—To describe an ellipse when the two axes are given.

In Fig. 5-49, draw the major and minor axes AB and CD, respectively, at right angles intersecting at E. On C, with AE as the radius, intersect the axis AB at F and G, the *foci;* insert pins through the axis at F and G, and loop a thread or cord on them equal in length to the axis AB, so that when stretched, it reaches extremity C of the *conjugate axis,* as shown in dotted lines. Place a pencil inside the cord, as at H, and, by guiding the pencil in this manner, describe the ellipse.

Second Method—Along the edge of a piece of paper, mark off a distance ac equal to AC, one-half the major axis, and from the same point a distance ab equal to CD, one-half the minor axis, as shown in Fig. 5-50. Place the paper so as to bring point b on the line AB, or major axis, and point c on the line DE, or minor axis. Lay off the position of point a. By shifting the paper so that point b travels on the major axis and point c travels on the minor axis, any

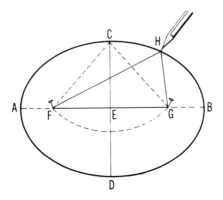

Fig. 5-49. To describe an ellipse when the two axes are given.

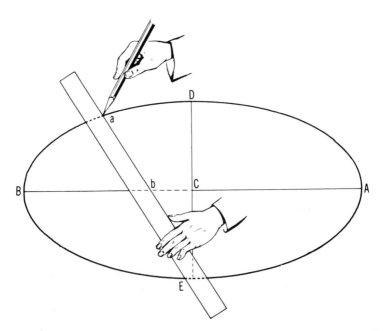

Fig. 5-50. To describe an ellipse given the two axes, second method.

number of points in the curve may be found through which the
curve may be traced.

Mensuration

As mentioned earlier, mensuration is the act, art, or process of
measuring. It is that branch of mathematics that deals with finding
the length of lines, the area of surfaces, and the volume of solids.
Therefore, the problems that follow will be divided into three
groups, as:

1. Measurement of lines, one dimension—length.
2. Measurement of surfaces (areas), two dimensions—length and
   width.
3. Measurement of solids (volumes), three dimensions—length,
   width, and thickness.

TABLE 5-3. Decimals of a Foot and Inches

Inch	0″	1″	2″	3″	4″	5″	6″	7″	8″	9″	10″	11″
0	.0000	.0833	.1677	.2500	.3333	.4167	.5000	.5833	.6667	.7500	.8333	.9167
1-16	.0052	.0885	.1719	.2552	.3385	.4219	.5052	.5885	.6719	.7552	.8385	.9219
1-8	.0104	.0937	.1771	.2604	.3437	.4271	.5104	.5937	.6771	.7604	.8437	.9271
3-16	.0156	.0990	.1823	.2656	.3490	.4323	.5156	.5990	.6823	.7656	.8490	.9323
1-4	.0208	.1042	.1875	.2708	.3542	.4375	.5208	.6042	.6875	.7708	.8542	.9375
5-16	.0260	.1094	.1927	.2760	.3594	.4427	.5260	.6094	.6927	.7760	.8594	.9427
3-8	.0312	.1146	.1979	.2812	.3646	.4479	.5312	.6146	.6979	.7812	.8646	.9479
7-16	.0365	.1198	.2031	.2865	.3698	.4531	.5365	.6198	.7031	.7865	.8698	.9531
1-2	.0417	.1250	.2083	.2917	.3750	.4583	.5417	.6250	.7083	.7917	.8750	.9583
9-16	.0469	.1302	.2135	.2969	.3802	.4635	.5469	.6302	.7135	.7969	.8802	.9635
5-8	.0521	.1354	.2188	.3021	.3854	.4688	.5521	.6354	.7188	.8021	.8854	.9688
11-16	.0573	.1406	.2240	.3073	.3906	.4740	.5573	.6406	.7240	.8073	.8906	.9740
3-4	.0625	.1458	.2292	.3125	.3958	.4792	.5625	.6458	.7292	.8125	.8958	.9792
13-16	.0677	.1510	.2344	.3177	.4010	.4844	.5677	.6510	.7344	.8177	.9010	.9844
7-8	.0729	.1562	.2396	.3229	.4062	.4896	.5729	.6562	.7396	.8229	.9062	.9896
15-16	.0781	.1615	.2448	.3281	.4115	.4948	.5781	.6615	.7448	.8281	.9115	.9948

Measurement of Lines—Length.

Problem 1—To find the length of any side of a right triangle given the other two sides.

Rule: The length of the hypotenuse equals the square root of the sum of the squares of the two legs; the length of either leg equals the square root of the difference of the square of the hypotenuse and the square of the other leg.

Example—The two legs of a right triangle measure 3 feet and 4 feet. Find the length of the hypotenuse. If the length of the hypotenuse and one leg are 5 feet and 4 feet, respectively, what is the length of the other leg?

In Fig. 5–51A,

$$AB = \sqrt{3^2 + 4^2} = \sqrt{25} = 5 \text{ feet}$$

In Fig. 5–51B,

$$BC = \sqrt{5^2 - 3^2} = \sqrt{25 - 9} = \sqrt{16} = 4 \text{ feet}$$

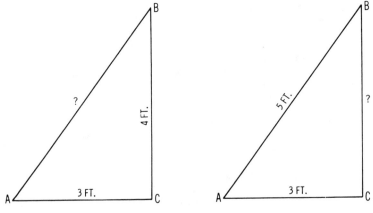

Fig. 5–51. To find the length of any side of a right triangle given the other two sides.

Problem 2—To find the length of the circumference of a circle.

Rule: Multiply the diameter by 3.1416.

Example—What length of molding strip is required for a circular window that is 5 feet in diameter?

$$5 \times 3.1416 = 15.7 \text{ feet}$$

Since the carpenter does not ordinarily measure feet in tenths, .7 should be reduced to inches; it corresponds to 8½ inches from Table 5–3. That is, the length of molding required is 15 feet 8½ inches.

Problem 3—To find the length of the arc of a circle.

Rule: Arc = .017453 × radius × central angle.

Example—If the radius of a circle is 2 feet, what is the length of a 60° arc?

Solution:
$$2 \times .017453 \times 60 = 2.094, \text{ or approximately 2 feet } 1\frac{1}{8} \text{ inches}$$

Problem 4—To find the rise of an arc.

Rule: Rise of an arc $= \sqrt{(4 \times \text{radius}^2) - \text{length}}$

Example—If the radius of a circle is 2 feet, what is the rise at the center of a 2-foot chord?

Solution:

$$\frac{1}{2}\sqrt{(4 \times 2^2) - 2} = \frac{1}{2}\sqrt{14} = 1.87 \text{ feet} = 1 \text{ foot } 10\frac{1}{2} \text{ inches}$$

Measurement of Surfaces—Area.

Problem 5—To find the area of a square.

Rule: Multiply the base by the height.

Example—What is the area of a square whose side is 5 feet (Fig. 5–52)?

$$5 \times 5 = 25 \text{ square feet}$$

Problem 6—To find the area of a rectangle.

Rule: Multiply the base by the height (i.e., width by length).

Example—What is the floor area of a porch 5 feet wide and 12 feet long (Fig. 5–53)?

$$5 \times 12 = 60 \text{ square feet}$$

Problem 7—To find the area of a parallelogram.

Rule: Multiply the base by the perpendicular height.

Example—What is the area of a 5′ × 12′ parallelogram (Fig. 5–54)?

$$5 \times 12 = 60 \text{ square feet}$$

Problem 8—To find the area of a triangle.

Rule: Multiply the base by one-half the altitude.

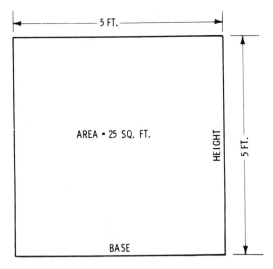

Fig. 5-52. To find the area of a square.

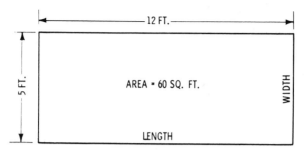

Fig. 5-53.   To find the area of a rectangle.

Example—How many square feet of sheathing are required to cover a church steeple having four triangular sides?

Problem 9—To find the area of a trapezoid.

Rule: Multiply one-half the sum of the two parallel sides by the perpendicular distance between them.

Example—What is the area of the trapezoid shown in Fig. 5-56?

LA and *FR* are the parallel sides, and *MS* is the perpendicular distance between them. Therefore,

$$\text{area} = \tfrac{1}{2}\,(LA + FR) \times MS$$

$$\text{area} = \tfrac{1}{2}\,(8 + 12) \times 6 = 60 \text{ square feet}$$

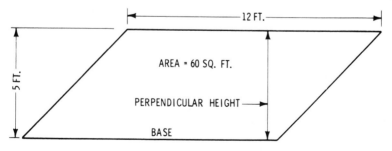

Fig. 5-54.   To find the area of a parallelogram.

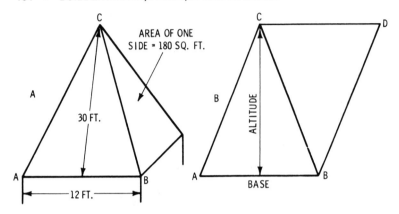

Fig. 5-55. To find the area of a triangle (equal to ½ area of parallelogram ABDC).

Problem 10—To find the area of a trapezium.

Rule: Draw a diagonal, dividing the figure into triangles; measure the diagonal and the altitudes, and find the area of the triangles; the sum of these areas is then the area of the trapezium.

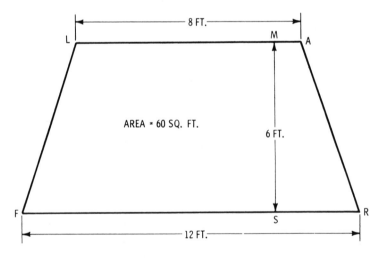

Fig. 5-56. To find the area of a trapezoid.

Example—What is the area of the trapezium shown in Fig. 5-57? (Draw diagonal *LR* and altitudes *AM* and *FS*.)

$$\text{area of triangle } ALR = \tfrac{1}{2}\,(12 \times 9) = 54 \text{ square feet}$$

$$\text{area of triangle } LRF = \tfrac{1}{2}\,(12 \times 6) = 36 \text{ square feet}$$

$$\text{area of trapezium } LARF = ALR + LRF = 36 + 54$$
$$= 90 \text{ square feet}$$

Problem 11—To find the area of any irregular polygon.

Rule: Draw diagonals, dividing the figure into triangles, and find the sum of the areas of these triangles.

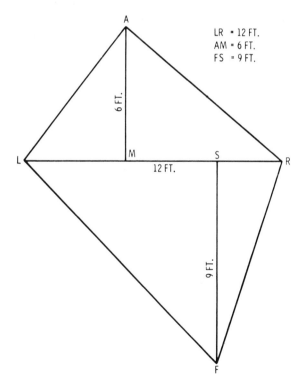

LR = 12 FT.
AM = 6 FT.
FS = 9 FT.

Fig. 5-57.   To find the area of a trapezium.

TABLE 5-4. Regular Polygons

Number of sides	3	4	5	6	7	8	9	10	11	12
Area when side = 1	.433	1.0	1.721	2.598	3.634	4.828	6.181	7.694	9.366	11.196

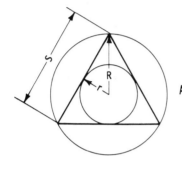

Steel square cut to miter sides at 30^0

A

$R = 2r$, or $.577$ S
$$AREA = .433 \ S^2$$
$$= 1.299 \ R^2$$
$$= 5.196 \ r^2$$
$$= \frac{(R + r)}{2} \ S$$

A. Equilateral triangle (3 sides).

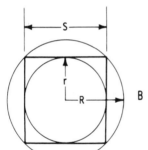

Steel square cut to miter sides at 45^0

B

$R = .707$ S, or $1.414 \ r$
$$AREA = S^2$$
$$= 2R^2$$
$$= 4r^2$$

B. Square (4 sides).

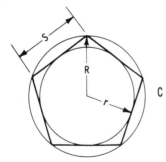

Steel square cut to miter sides at 36^0

C

$R = .851$ S, or $1.236 \ r$
$$AREA = 1.72 \ S^2$$
$$= 2.378 \ R^2$$
$$= 3.633 \ r^2$$

C. Pentagon (5 sides).

Fig. 5-58. Regular polygons.

Problem 12—To find the area of any regular polygon, such as shown in Fig. 5–58, when the length of only one side is given.

Rule: Multiply the square of the sides by the figure for "area when side = 1" opposite the particular polygon in Table 5–4.

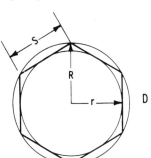

Steel square cut to miter sides at 30°

R = S, or 1.155 r
AREA = 2.598 S²
     = 2.598 R²
     = 3.464 r²

D. Hexagon (6 sides).

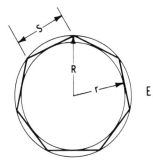

Steel square cut to miter corners at 25° 43'

R = 1.152 S, or 1.11 r
AREA = 3.634 S²
     = 2.736 R²
     = 3.371 r²

E. Heptagon (7 sides).

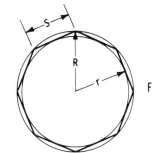

Steel square cut to miter corners at 22° 30'

R = 1.307S or 1.082 r
AREA = 4.828 S²
     = 2.828 R²
     = 3.314 r²

F. Octagon (8 sides).

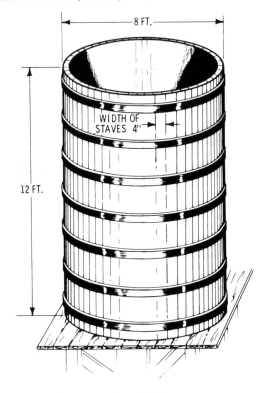

Fig. 5-60.   To find the area of a cylinder.

$$\text{cylindrical surface} = 3.1416 \times 8 \times 12 = 302 \text{ square feet}$$

$$\text{circumference of tank} = 3.1416 \times 8 = 25.1 \text{ feet}$$

$$\text{number of } 4'' \times 12' \text{ pieces} = \frac{25.1 \times 12}{4} = 25.1 \times 3 = 75.3$$

Problem 19—To find the area of a cone (Fig. 5–61).

Rule: Multiply 3.1416 by the diameter of the base and by one-half the slant height.

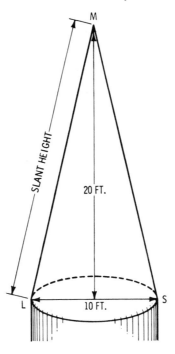

Fig. 5-61. To find the surface area of a cone.

Example—A conical spire with a base 10 feet in diameter and an altitude of 20 feet is to be covered. Find the area of the surface to be covered.

$$\text{slant height} = \sqrt{5^2 + 20^2} = \sqrt{425} = 20.62 \text{ feet}$$

$$\text{circumference of base} = 3.1416 \times 10 = 31.416 \text{ feet}$$

$$\text{area of conical surface} = 31.416 \times \tfrac{1}{2} \times 20.62 = 324 \text{ square feet}$$

Problem 20—To find the area of the frustum of a cone (Fig. 5–62).

Rule: Multiply one-half the slant height by the sum of the circumference.

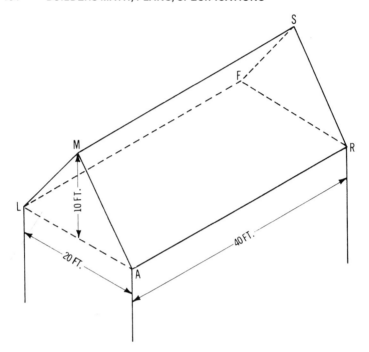

Fig. 5-63. To find the volume of a rectangular wedge.

required parts of triangles from given parts. The only branch of trigonometry useful to the carpenter and builder is plane trigonometry, where the lines in the triangles are straight and where they all lie in the same plane.

There are six elements, or parts, in every triangle—three sides and three angles. The sum of the three angles, no matter what the lengths of the sides, will always be equal to 180 degrees.

When any three of the six parts are given, provided one or more of them are sides, the other three are calculable. The angles are measured in circular measure—in degrees (°), minutes ('), and seconds ("). The term *degree* has no numerical value; in trigonometry it simply means $\frac{1}{360}$ of a circle, nothing more.

To the student of trigonometry, any two radii that divide a circle into anything more than 0° or less than 360° form an angle.

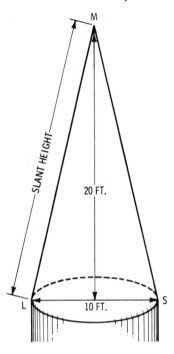

Fig. 5-61. To find the surface area of a cone.

Example—A conical spire with a base 10 feet in diameter and an altitude of 20 feet is to be covered. Find the area of the surface to be covered.

$$\text{slant height} = \sqrt{5^2 + 20^2} = \sqrt{425} = 20.62 \text{ feet}$$

$$\text{circumference of base} = 3.1416 \times 10 = 31.416 \text{ feet}$$

$$\text{area of conical surface} = 31.416 \times \tfrac{1}{2} \times 20.62 = 324 \text{ square feet}$$

Problem 20—To find the area of the frustum of a cone (Fig. 5–62).

Rule: Multiply one-half the slant height by the sum of the circumference.

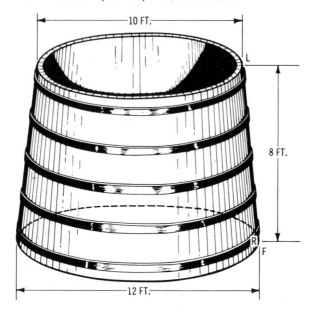

Fig. 5-62. To find the area of the frustum of a cone.

Example—A tank is 12 feet in diameter at the base, 10 feet at the top, and 8 feet high. What is the area of the slant surface?

circumference of 10-foot diameter = 3.1416 × 10 = 31.416 feet

circumference of 12-foot diameter = 3.1416 × 12 = 37.7 feet

sum of circumferences = 69.1 feet

slant height = $\sqrt{1^2 + 8^2}$ = $\sqrt{65}$ = 8.12

slant surface = sum of circumferences × ½ slant height

slant surface = 69.1 × ½ × 8.12 = 280 square feet

Measurement of Solids—Volume

Problem 21—To find the volume of a rectangular solid.

Rule: Multiply the length, width, and thickness together.

Example—What is the volume of a 4″ × 8″ × 12′ timber? (Before applying the rule, reduce all dimensions to feet.)

$$4'' = \frac{1}{3} \text{ foot}$$

$$8'' = \frac{2}{3} \text{ foot}$$

volume of timber = $\frac{1}{3} \times \frac{2}{3} \times 12 = 2.67$ cubic feet

If the timber were a piece of oak weighing 48 pounds per cubic foot, the total weight would be

$$48 \times 2.67 = 128 \text{ pounds}$$

Problem 22—To find the volume of a rectangular wedge.

Rule: Find the area of one of the triangular ends, and multiply the area by the distance between the ends.

Example—An attic has the shape of a rectangular wedge. What volume storage capacity would there be for the proportions shown in Fig. 5–63? In the illustration, the boundary of the attic is *LARFMS*.

area of triangular end *MLA* = $20 \times \frac{10}{2} = 100$ square feet

volume of attic = $100 \times 40 = 4000$ cubic feet

Trigonometry

Trigonometry is that branch of mathematics that deals with the relations that exist between the sides and angles of triangles, and more especially with those of the methods of calculating the

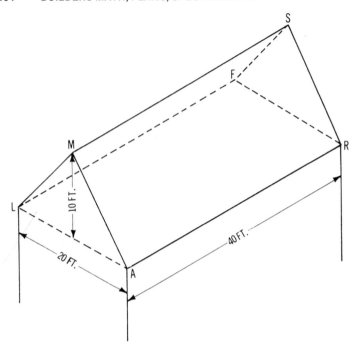

Fig. 5-63. To find the volume of a rectangular wedge.

required parts of triangles from given parts. The only branch of trigonometry useful to the carpenter and builder is plane trigonometry, where the lines in the triangles are straight and where they all lie in the same plane.

There are six elements, or parts, in every triangle—three sides and three angles. The sum of the three angles, no matter what the lengths of the sides, will always be equal to 180 degrees.

When any three of the six parts are given, provided one or more of them are sides, the other three are calculable. The angles are measured in circular measure—in degrees (°), minutes ('), and seconds ("). The term *degree* has no numerical value; in trigonometry it simply means $\frac{1}{360}$ of a circle, nothing more.

To the student of trigonometry, any two radii that divide a circle into anything more than 0° or less than 360° form an angle.

The first 90° division is called the *first quadrant*. Angles in this quadrant are the *acute angles* (Fig. 5–64A) mentioned earlier in this chapter. Angles from 90° to 180° are in the *second quadrant*. These are the *obtuse angles* (Fig. 5–64B) mentioned. Angles from 180° to 270° lie in the *third quadrant*, and angles from 270° to 360° lie in the *fourth quadrant*. These quadrants are represented pictorially in Fig. 5–65. Only angles in the first and second quadrants, from 0° to 180°, will be discussed in this section. Note that a straight line may be considered as an angle of 180°. Trigonometry is actually based on geometry, but it makes use of many algebraic operations that can be used by carpenters and builders.

Trigonometric Functions

In mathematics, a *function* means a quantity that necessarily changes because of a change in another number with which it is connected in some way. In trigonometry, it is probably less confusing to call the trigonometric functions simply *ratios*, which they truly are.

Refer to Fig. 5–66 for an explanation of the following. There are six trigonometric ratios commonly used:

$$\text{Sine of angle } A = \frac{\text{opposite side}}{\text{hypotenuse}} \text{ or } \frac{BC}{AB}$$

$$\text{Cosine of angle } A = \frac{\text{adjacent side}}{\text{hypotenuse}} \text{ or } \frac{AC}{AB}$$

$$\text{Tangent of angle } A = \frac{\text{opposite side}}{\text{adjacent side}} \text{ or } \frac{BC}{AC}$$

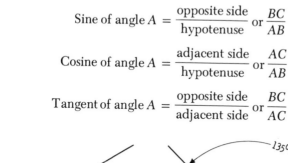

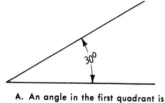

A. An angle in the first quadrant is an acute angle.

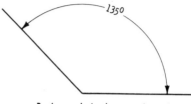

B. An angle in the second quadrant is an obtuse angle.

Fig. 5-64. Acute and obtuse angles.

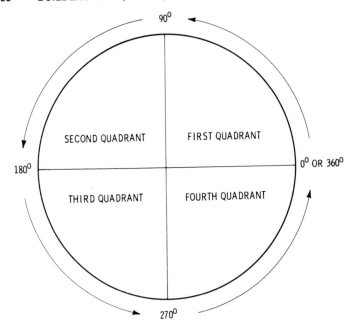

Fig. 5–65. The four quadrants of a circle.

$$\text{Cotangent of angle } A = \frac{\text{adjacent side}}{\text{opposite side}} \text{ or } \frac{AC}{BC}$$

$$\text{Secant of angle } A = \frac{\text{hypotenuse}}{\text{adjacent side}} \text{ or } \frac{AB}{AC}$$

$$\text{Cosecant of angle } A = \frac{\text{hypotenuse}}{\text{opposite side}} \text{ or } \frac{AB}{BC}$$

Note that these last three functions are only *reciprocals* of the sine, cosine, and tangent, respectively, or

$$\text{cosecant} = \frac{1}{\text{sine}}$$

$$\text{secant} = \frac{1}{\text{cosine}}$$

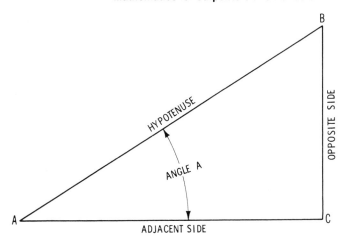

Fig. 5-66. A right triangle illustrates the application of trigonometric ratios which are commonly used.

$$\text{cotangent} = \frac{1}{\text{tangent}}$$

If a proposition calls for multiplication by the sine of an angle, the same result will be obtained by dividing by the cosecant. It is convenient to do this in many calculations.

It is impossible in a discussion of this type to give a comprehensive table of the trigonometric ratios, although an adequate but limited number of trigonometric functions are presented in Table 5-5. Anyone who would like to follow up the information given here is advised to obtain a book of five- or six-place tables.

As an example of how trigonometric ratios are used to solve one of the carpenter's most common problems, determining the length of rafters given the rise and run, refer to Fig. 5-67. The slope of the roof, in degrees, may be determined by dividing the opposite side 12 feet, by the adjacent side, 18 feet. This is the tangent of the angle A and is equal to $^{12}\!/_{18}$, or .6667. From Table 5-5, angle A is determined to be 33° 42′. The length of the rafter may be determined by the ratio:

$$\text{secant} = \frac{\text{hypotenuse}}{\text{adjacent side}}$$

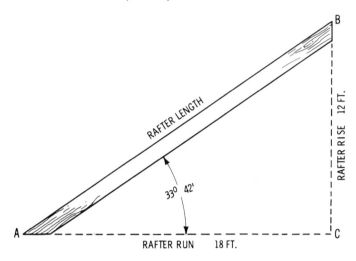

Fig. 5-67. Trigonometric ratios may be used to determine the lengths of rafters for a roof.

or the hypotenuse (the length of the rafter) is equal to

$$secant\ 33°42' \times adjacent\ side$$

The secant of 33°42' is equal to 1.2020. Therefore, the calculation for the length of the rafter is

$$1.2020 \times 18 = 21.64\ feet = 21\ feet\ 7\tfrac{3}{16}\ inches$$

Since the opposite side is known to be 12 feet, the calculation could just as easily be made by using the cosecant function.

The slopes, in degrees, for all regular roof pitches are given in Table 5-6; these pitches range from 12 × 1 to 12 × 12, and the three main trigonometric ratios—sine, cosine, and tangent—are provided for each pitch.

Other typical examples of how trigonometric ratios can aid carpenters are shown in the following problems.

Problem 1—A grillwork consisting of radial and vertical members is to be built in a semicircular opening with a radius of

TABLE 5-5. Natural Trigonometric Functions

Degree	Sine	Cosine	Tangent	Secant	Degree	Sine	Cosine	Tangent	Secant
0	.00000	1.0000	.00000	1.0999	46	.7193	.6947	1.0355	1.4395
1	.01745	.9998	.01745	1.0001	47	.7314	.6820	1.0724	1.4663
2	.03490	.9994	.03492	1.0006	48	.7431	.6691	1.1106	1.4945
3	.05234	.9986	.05241	1.0014	49	.7547	.6561	1.1504	1.5242
4	.06976	.9976	.06993	1.0024	50	.7660	.6428	1.1918	1.5557
5	.08716	.9962	.08749	1.0038	51	.7771	.6293	1.2349	1.5890
6	.10453	.9945	.10510	1.0055	52	.7880	.6157	1.2799	1.6243
7	.12187	.9925	.12278	1.0075	53	.7986	.6018	1.3270	1.6616
8	.1392	.9903	.1405	1.0098	54	.8090	.5878	1.3764	1.7013
9	.1564	.9877	.1584	1.0125	55	.8192	.5736	1.4281	1.7434
10	.1736	.9848	.1763	1.0154	56	.8290	.5592	1.4826	1.7883
11	.1908	.9816	.1944	1.0187	57	.8387	.5446	1.5399	1.8361
12	.2079	.9781	.2126	1.0223	58	.8480	.5299	1.6003	1.8871
13	.2250	.9744	.2309	1.0263	59	.8572	.5150	1.6643	1.9416
14	.2419	.9703	.2493	1.0306	60	.8660	.5000	1.7321	2.0000
15	.2588	.9659	.2679	1.0353	61	.8746	.4848	1.8040	2.0627
16	.2756	.9613	.2867	1.0403	62	.8829	.4695	1.8807	2.1300
17	.2924	.9563	.3057	1.0457	63	.8910	.4540	1.9626	2.2027
18	.3090	.9511	.3249	1.0515	64	.8988	.4384	2.0503	2.2812
19	.3256	.9455	.3443	1.0576	65	.9063	.4226	2.1445	2.3662
20	.3420	.9397	.3640	1.0642	66	.9135	.4067	2.2460	2.4586
21	.3584	.9336	.3839	1.0711	67	.9205	.3907	2.3559	2.5598
22	.3746	.9272	.4040	1.0785	68	.9272	.3746	2.4751	2.6695
23	.3907	.9205	.4245	1.0864	69	.9336	.3584	2.6051	2.7904
24	.4067	.9135	.4452	1.0946	70	.9397	.3420	2.7475	2.9238
25	.4226	.9063	.4663	1.1034	71	.9455	.3256	2.9042	3.0715
26	.4384	.8988	.4877	1.1126	72	.9511	.3090	3.0777	3.2361
27	.4540	.8910	.5095	1.1223	73	.9563	.2924	3.2709	3.4203
28	.4695	.8829	.5317	1.1326	74	.9613	.2756	3.4874	3.6279
29	.4848	.8746	.5543	1.1433	75	.9659	.2588	3.7321	3.8637
30	.5000	.8660	.5774	1.1547	76	.9703	.2419	4.0108	4.1336
31	.5150	.8572	.6009	1.1663	77	.9744	.2250	4.3315	4.4454
32	.5299	.8480	.6249	1.1792	78	.9781	.2079	4.7046	4.8097
33	.5446	.8387	.6494	1.1924	79	.9816	.1908	5.1446	5.2408
34	.5592	.8290	.6745	1.2062	80	.9848	.1736	5.6713	5.7588
35	.5736	.8192	.7002	1.2208	81	.9877	.1564	6.3138	6.3924
36	.5878	.8090	.7265	1.2361	82	.9903	.1392	7.1154	7.1853
37	.6018	.7986	.7536	1.2521	83	.9925	.12187	8.1443	8.2055
38	.6157	.7880	.7813	1.2690	84	.9945	.10453	9.5144	9.5668
39	.6293	.7771	.8098	1.2867	85	.9962	.08716	11.4301	11.474
40	.6428	.7660	.8391	1.3054	86	.9976	.06976	14.3007	14.335
41	.6561	.7547	.8693	1.3250	87	.9986	.05234	19.0811	19.107
42	.6691	.7431	.9004	1.3456	88	.9994	.03490	28.6363	28.654
43	.6820	.7314	.9325	1.3673	89	.9998	.01745	57.2900	57.299
44	.6947	.7193	.9657	1.3902	90	1.0000	Inf.	Inf.	Inf.
45	.7071	.7071	1.0000	1.4142	—	—	—	—	—

TABLE 5-6. Roof Pitches in Degrees and Minutes
(Measured from the horizontal)

Pitch	Sine	Cosine	Tangent
12 × 1 = 4° 46′	.083098	.996541	.083386
12 × 2 = 9° 28′	.164474	.986381	.166745
12 × 3 = 14° 2′	.242486	.970155	.249946
12 × 4 = 18° 26′	.316201	.948692	.333302
12 × 5 = 22° 37′	.384564	.923098	.416601
12 × 6 = 26° 34′	.444635	.895712	.496404
12 × 7 = 30° 15′	.503774	.863836	.583183
12 × 8 = 33° 41′	.554602	.832115	.666497
12 × 9 = 36° 53′	.600188	.799859	.750366
12 × 10 = 39° 46′	.639663	.768656	.832183
12 × 11 = 42° 31′	.675805	.737081	.916866
12 × 12 = 45° 00′	.707107	.707107	1.000000

6 feet, as shown in Fig. 5–68. Find the lengths of the vertical pieces MS and LF.

For triangle OMS, the hypotenuse is known to be 6 feet, and angle O is 30°. Line MS is the opposite side of the triangle, and

$$\text{opposite side} = 30°, \text{or}$$

$$\frac{\text{opposite side}}{\text{hypotenuse}} = \text{sine } 30° \times \text{hypotenuse}$$

$$\text{sine } 30° = .500$$

This is the calculation:

$$\text{opposite side} = .500 \times 6 = 3 \text{ feet}$$

For triangle OLF, the hypotenuse is 6 feet, and angle O is 60°.

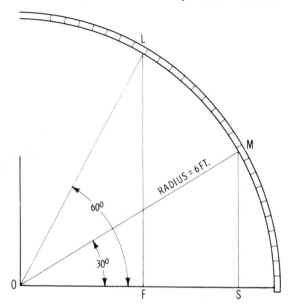

Fig. 5-68. The method of finding the length of vertical pieces in grillwork with the aid of trigonometric relations.

Line *LF* is the opposite side, and

$$\frac{\text{opposite side}}{\text{hypotenuse}} = 60°,$$

or opposite side = 60° × hypotenuse.

$$\text{sine } 60° = .866$$

This is the calculation:

opposite side = .866 × 6 = 5.196 feet = 5 feet 2⅜ inches

Problem 2—When laying out the grillwork in Fig. 5–68, how far must the members *LF* and *MS* be spaced from the center *O* to be vertical?

The hypotenuse is known to be 6 feet and the length of adjacent side OF is to be found.

$$\frac{\text{adjacent side}}{\text{hypotenuse}} = \cos 60°, \text{ or}$$

$$\text{adjacent side} = \cos 60° \times \text{hypotenuse}.$$

$$\cos 60° = .500$$

This is the calculation:

$$\text{adjacent side } OF = .500 \times 6 = 3 \text{ feet}$$

For the length of the adjacent side OS,

$$\frac{\text{adjacent side}}{\text{hypotenuse}} = \cos 30°,$$

$$\text{or adjacent side} = \cos 30° \times \text{hypotenuse}.$$

$$(\cos 30° = .866)$$

This is the calculation:

$$\text{adjacent side } OS = .866 \times 6 = 5.196 \text{ feet} = 5 \text{ feet } 2\frac{3}{8} \text{ inches}$$

Problem 3—A bridge is to be constructed from the top of a building to an opening in the roof of an adjacent building, as in Fig. 5–69. If the rise OF to the point of entry L is 15 feet and the pitch of the roof is $\frac{1}{2}$, what length beams FL are required?

From Table 5–6, $\frac{1}{2}$ pitch, or $12'' \times 12''$, is 45°. The adjacent side OF is known to be 15 feet. The required length of the opposite side = adjacent side $\times$ tan 45°. This is the calculation:

$$\text{opposite side } FL = 15 \times 1.00 = 15 \text{ feet}$$

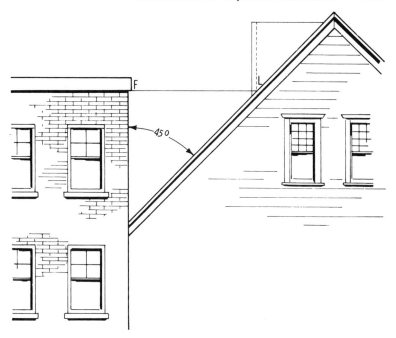

Fig. 5-69. The method of finding the distance from one side of a building to a given point on an adjacent roof by employing trigonometric relations.

Problem 4—When estimating the amount of roofing material necessary to cover the side of the roof from O to L in Fig. 5-69, what is the distance from O to L?

It is required to find the hypotenuse with the adjacent side and included angle being given.

$$\frac{\text{hypotenuse}}{\text{adjacent side}} = \text{secant angle } 0, \text{ or}$$

hypotenuse $= 1.4142 \times 15 = 21.213$ feet $= 21$ feet $2\frac{9}{16}$ inches.

TABLE 5-7. Functions of Numbers

No.	Square	Cube	Square Root	Cubic Root	Logarithm	1000 × Reciprocal	No. = Diameter Circum.	Area
1	1	1	1.0000	1.0000	0.00000	1000.000	3.142	0.7854
2	4	8	1.4142	1.2599	0.30103	500.000	6.283	3.1416
3	9	27	1.7321	1.4422	0.47712	333.333	9.425	7.0686
4	16	64	2.0000	1.5874	0.60206	250.000	12.566	12.5664
5	25	125	2.2361	1.7100	0.69897	200.000	15.708	19.6350
6	36	216	2.4495	1.8171	0.77815	166.667	18.850	28.2743
7	49	343	2.6458	1.9129	0.84510	142.857	21.991	38.4845
8	64	512	2.8284	2.0000	0.90309	125.000	25.133	50.2655
9	81	729	3.0000	2.0801	0.95424	111.111	28.274	63.6173
10	100	1000	3.1623	2.1544	1.00000	100.000	31.416	78.5398
11	121	1331	3.3166	2.2240	1.04139	90.9091	34.558	95.0332
12	144	1728	3.4641	2.2894	1.07918	83.3333	37.699	113.097
13	169	2197	3.6056	2.3513	1.11394	76.9231	40.841	132.732
14	196	2744	3.7417	2.4101	1.14613	71.4286	43.982	153.938
15	225	3375	3.8730	2.4662	1.17609	66.6667	47.124	176.715
16	256	4096	4.0000	2.5198	1.20412	62.5000	50.265	201.062
17	289	4913	4.1231	2.5713	1.23045	58.8235	53.407	226.980
18	324	5832	4.2426	2.6207	1.25527	55.5556	56.549	254.469
19	361	6859	4.3589	2.6684	1.27875	52.6316	59.690	283.529
20	400	8000	4.4721	2.7144	1.30103	50.0000	62.832	314.159
21	441	9261	4.5826	2.7589	1.32222	47.6190	65.973	346.361
22	484	10648	4.6904	2.8020	1.34242	45.4545	69.115	380.133
23	529	12167	4.7958	2.8439	1.36173	43.4783	72.257	415.476
24	576	13824	4.8990	2.8845	1.38021	41.6667	75.398	452.389
25	625	15625	5.0000	2.9240	1.39794	40.0000	78.540	490.874
26	676	17576	5.0990	2.9625	1.41497	38.4615	81.681	530.929
27	729	19683	5.1962	3.0000	1.43136	37.0370	84.823	572.555
28	784	21952	5.2915	3.0366	1.44716	35.7143	87.965	615.752
29	841	24389	5.3852	3.0723	1.46240	34.4828	91.106	660.520
30	900	27000	5.4772	3.1072	1.47712	33.3333	94.248	706.858
31	961	29791	5.5678	3.1414	1.49136	32.2581	97.389	754.768
32	1024	32768	5.6569	3.1748	1.50515	31.2500	100.531	804.248
33	1089	35937	5.7446	3.2075	1.51851	30.3030	103.673	855.299
34	1156	39304	5.8310	3.2396	1.53148	29.4118	106.814	907.920
35	1225	42875	5.9161	3.2711	1.54407	28.5714	109.956	962.113
36	1296	46656	6.0000	3.3019	1.55630	27.7778	113.097	1017.88
37	1369	50653	6.0828	3.3322	1.56820	27.0270	116.239	1075.21
38	1444	54872	6.1644	3.3620	1.57978	26.3158	119.381	1134.11
39	1521	59319	6.2450	3.3912	1.59106	25.6410	122.522	1194.59
40	1600	64000	6.3246	3.4200	1.60206	25.0000	125.66	1256.64
41	1681	68921	6.4031	3.4482	1.61278	24.3902	128.81	1320.25
42	1764	74088	6.4807	3.4760	1.62325	23.8095	131.95	1385.44
43	1849	79507	6.5574	3.5034	1.63347	23.2558	135.09	1452.20
44	1936	85184	6.6332	3.5303	1.64345	22.7273	138.23	1520.53
45	2025	91125	6.7082	3.5569	1.65321	22.2222	141.37	1590.43

TABLE 5-7. Functions of Numbers (Continued)

No.	Square	Cube	Square Root	Cubic Root	Logarithm	1000 × Reciprocal	No. = Diameter Circum.	Area
46	2116	97336	6.7823	3.5830	1.66276	21.7391	144.51	1661.90
47	2209	103823	6.8557	3.6088	1.67210	21.2766	147.65	1734.94
48	2304	110592	6.9282	3.6342	1.68124	20.8333	150.80	1809.56
49	2401	117649	7.0000	3.6593	1.69020	20.4082	153.94	1885.74
50	2500	125000	7.0711	3.6840	1.69897	20.0000	157.08	1963.50
51	2601	132651	7.1414	3.7084	1.70757	19.6078	160.22	2042.82
52	2704	140608	7.2111	3.7325	1.71600	19.2308	163.36	2123.72
53	2809	148877	7.2801	3.7563	1.72428	18.8679	166.50	2206.18
54	2916	157464	7.3485	3.7798	1.73239	18.5185	169.65	2290.22
55	3025	166375	7.4162	3.8030	1.74036	18.1818	172.79	2375.83
56	3136	175616	7.4833	3.8259	1.74819	17.8571	175.93	2463.01
57	3249	185193	7.5498	3.8485	1.75587	17.5439	179.07	2551.76
58	3364	195112	7.6158	3.8709	1.76343	17.2414	182.21	2642.08
59	3481	205379	7.6811	3.8930	1.77085	16.9492	185.35	2733.97
60	3600	216000	7.7460	3.9149	1.77815	16.6667	188.50	2827.43
61	3721	226981	7.8102	3.9365	1.78533	16.3934	191.64	2922.47
62	3844	238328	7.8740	3.9579	1.79239	16.1290	194.78	3019.07
63	3969	250047	7.9373	3.9791	1.79934	15.8730	197.92	3117.25
64	4096	262144	8.0000	4.0000	1.80618	15.6250	201.06	3216.99
65	4225	274625	8.0623	4.0207	1.81291	15.3846	204.20	3318.31
66	4356	287496	8.1240	4.0412	1.81954	15.1515	207.35	3421.19
67	4489	300763	8.1854	4.0615	1.82607	14.9254	210.49	3525.65
68	4624	314432	8.2462	4.0817	1.83251	14.7059	213.63	3631.68
69	4761	328509	8.3066	4.1016	1.83885	14.4928	216.77	3739.28
70	4900	343000	8.3666	4.1213	1.84510	14.2857	219.91	3848.45
71	5041	357911	8.4261	4.1408	1.85126	14.0845	223.05	3959.19
72	5184	373248	8.4853	4.1602	1.85733	13.8889	226.19	4071.50
73	5329	389017	8.5440	4.1793	1.86332	13.6986	229.34	4185.39
74	5476	405224	8.6023	4.1983	1.86923	13.5135	232.48	4300.84
75	5625	421875	8.6603	4.2172	1.87506	13.3333	235.62	4417.86
76	5776	438976	8.7178	4.2358	1.88081	13.1579	238.76	4536.46
77	5929	456533	8.7750	4.2543	1.88649	12.9870	241.90	4656.63
78	6084	474552	8.8318	4.2727	1.89209	12.8205	245.04	4778.36
79	6241	493039	8.8882	4.2908	1.89763	12.6582	248.19	4901.67
80	6400	512000	8.9443	4.3089	1.90309	12.5000	251.33	5026.55
81	6561	531441	9.0000	4.3267	1.90849	12.3457	254.47	5153.00
82	6724	551368	9.0554	4.3445	1.91381	12.1951	257.61	5281.02
83	6889	571787	9.1104	4.3621	1.91908	12.0482	260.75	5410.61
84	7056	592704	9.1652	4.3795	1.92428	11.9048	263.89	5541.77
85	7225	614125	9.2195	4.3968	1.92942	11.7647	267.04	5674.50
86	7396	636056	9.2736	4.4140	1.93450	11.6279	270.18	5808.80
87	7569	658503	9.3274	4.4310	1.93952	11.4943	273.32	5944.68
88	7744	681472	9.3808	4.4480	1.94448	11.3636	276.46	6082.12
89	7921	704969	9.4340	4.4647	1.94939	11.2360	279.60	6221.14

TABLE 5-7.   Functions of Numbers (Continued)

No.	Square	Cube	Square Root	Cubic Root	Logarithm	1000 × Reciprocal	No. = Diameter Circum.	No. = Diameter Area
90	8100	729000	9.4868	4.4814	1.95424	11.1111	282.74	6361.73
91	8281	753571	9.5394	4.4979	1.95904	10.9890	285.88	6503.88
92	8464	778688	9.5917	4.5144	1.96379	10.8696	289.03	6647.61
93	8649	804357	9.6437	4.5307	1.96848	10.7527	292.17	6792.91
94	8836	830584	9.6954	4.5468	1.97313	10.6383	295.31	6939.78
95	9025	857375	9.7468	4.5629	1.97772	10.5263	298.45	7088.22
96	9216	884736	9.7980	4.5789	1.98227	10.4167	301.59	7238.23
97	9409	912673	9.8489	4.5947	1.98677	10.3093	304.73	7389.81
98	9604	941192	9.8995	4.6104	1.99123	10.2041	307.88	7542.96
99	9801	970299	9.9499	4.6261	1.99564	10.1010	311.02	7697.69

Summary

All the sciences are based on arithmetic and the ability to use it. Arithmetic is the art of calculating by using numbers. A number is a total amount, or aggregate, of units. By computing the units, we arrive at a certain number or total. By the same token, a unit means a single article, often a definite group adopted as a standard of measurement, such as dozen, ton, foot, bushel, or mile.

Fractions indicate that a number or unit has been divided into a certain number of equal parts, and shows how many of these parts are to be considered. Two forms of fractions are in common usage—the decimal and the common fraction. The common fraction is written by using two numbers, one written over or alongside the other with a line between them, the lower (or second) number being called the denominator, and the upper (or first) number being called the numerator.

Geometry is a branch of mathematics that deals with space and figures in space. It is the science of the mutual relations of points, lines, angles, surfaces, and solids, which are considered as having no properties except those arising from extension and difference of situation. There are two kinds of lines—straight and curved. A straight line is the shortest distance between two points. A curved line is one that changes its direction at every point.

Trigonometry is the branch of mathematics that deals with the relations that exist between the sides and angles of triangles, especially the methods of calculating the required parts of triangles from given parts. There are six elements, or parts, in every triangle—three sides and three angles. The sum of the three angles, no matter what the lengths of the sides, will always be equal to 180 degrees.

Review Questions

1. What is the definition of arithmetic?
2. What are even numbers?
3. What are odd numbers?
4. What are fractions? How are they used?
5. What is trigonometry and how is it used in carpentry?

CHAPTER 6

Surveying

By definition, surveying means the art or science of determining the area and configuration of portions of the surface of the earth. There are two general divisions of surveying that may be classified with respect to the nature of the measurements taken as:

1. Leveling.
2. Measurement of angles (transit work).

Leveling, in surveying, is the operation of determining the comparative levels of different points of land for the purpose of laying out a grade or building site, etc., by sighting through a leveling instrument at one point to a leveling staff at another point, as shown in Fig. 6-1.

The Level

This instrument (Fig. 6-2), is employed to determine the difference in elevation between points. A common form is known as the wye level, so called because its shape resembles the letter Y. It consists of a telescope mounted on two supports which from their shape are called Y's. The crossbar supporting the telescope is attached to a vertical spindle, which allows it to be turned in a hori-

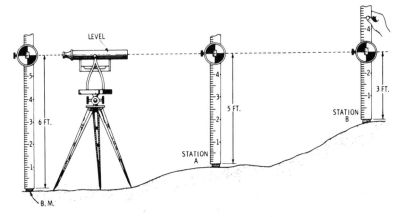

Fig. 6–1. The terms "backsight" and "foresight" do not necessarily mean backward and forward. Readings taken on a point of known elevation, such as a bench mark or a turning point, are called backsights, whereas readings taken on a point of unknown elevation are termed foresights. In the illustration, the backsight (to the bench mark) is +6 feet 0 inches, and the foresight to station A is −5 feet 0 inches. The difference in the two elevations is 6 feet − 5 feet = 1 foot. If the reading at station A had been greater than the 6-foot HI (height of instrument), the calculation would have shown a negative result, thereby indicating that station A was lower than the bench mark. The elevation at station B is calculated in the same manner: 6 − 3 = 3 feet. Therefore, the elevations from any number of points can be obtained in the same way, if they can be seen from the same position of the instrument; if they cannot, a new HI must be used.

zontal plane. Directly beneath the telescope and attached parallel to it is a spirit level by means of which the line of collimation of the telescope may be rendered horizontal. The line of collimation is the line that would connect the intersection of the cross hairs with the optical center of the objective.

Construction of the Wye Level

In construction, a circular plate is screwed to a tripod, and to this is attached a similar plate parallel to the first and connected with it by a ball-and socket joint. Four screws (sometimes only

Fig. 6-2. A typical builder's level. *(Courtesy David White Instruments, Division of Realist, Inc.)*

three), called foot or plate screws, hold these plates apart by resting on the lower one and passing through the other. A vertical spindle in the center of the plates supports a rod, bar, or beam and is used to revolve the instrument. The beam is horizontal and carries at its ends two vertical standards or supports of equal size terminated by two forks of the general form of the letter Y. The inside of the Y's is Y-shaped, with an open bottom to prevent an accumulation of dirt. The top of the Y's may be closed by semicircular straps or bridles called clips, which are hinged on one side and pinned on the other. The pins are tapered to permit fastening of the telescope. It is *never* clamped tightly.

The tops of the Y's and the corresponding clips are called the rings or collars and should be of equal diameter. A telescope is placed on the rings which support a spirit level. A clamping screw just above the upper plate serves to secure the instrument in any position desired. A tangent screw, also above the upper plate, provides slow-motion, or vernier, adjustment to the instrument.

The Telescope—The "objective," or "object glass" (so called because it faces the object looked at), is a compound lens that is made to correct spherical and chromatic aberrations of single lenses. It gathers light and forms an image at a point in the tube where cross hairs are placed. The ocular, or eye, piece is also a compound lens through which the operator looks to see a magnified view of the image. In the best precision instruments, often foreign-made, the image is often *inverted*. A good instrument man quickly becomes accustomed to the inverted image, but most American-made instruments have an *erecting* image, which shows the object right side up. Tangent screws may be used to give motion to the tubes carrying the objective and ocular.

The Cross Hairs—These are made of platinum-drawn wires or "spider's threads" attached to a ring within the telescope at the spot where the image is formed. The ring is secured by four capstan-headed screws which pass through the telescope tube. There are commonly two hairs, one horizontal and the other vertical, with their intersection in the axis of the telescope.

Bubble Level—The spirit level attached to the telescope can be raised vertically by means of altitude screws at the rear end, and it may be moved laterally to a limited extent by means of azimuth screws at the forward end.

The Supports—These form the Y's and are supported by the bar to which they are fastened by two nuts, one above and one below. These nuts may be moved to provide an adjustment, to move the scope in a horizontal direction.

Lines of the Level

There are three principal lines of a level:

1. Vertical axis.
2. Bubble line.
3. Line of collimation.

The Vertical Axis—This passes through the center of the spindle.

Bubble Line—The metallic supports of the spirit level are equal, and the tangent at their top or bottom is horizontal when the bubble is centered. This tangent is the bubble line.

Line of Collimation—The line that would connect the intersection of the cross hairs with the optical center of the objective is the line of collimation.

Relations Between the Lines of the Level—The following relations must be obtained:

1st Relation—The bubble line and the line of collimation must be parallel.

2nd Relation—The plane described by the bubble line should be horizontal, that is, perpendicular to the vertical axis. These conditions are generally satisfied in a new level, but exposure and use may alter these relations; therefore there is the necessity of adjusting the instrument occasionally.

Adjustments of the Wye Level

Levels and transits are expensive, precision instruments and should be treated as such. Although a passable job of leveling may be done by a relatively inexperienced man, it is questionable if a major job of adjusting should be attempted by a novice. A perfect job of adjustment is difficult, even for an experienced adjuster, and there are few instruments in perfect adjustment. For precision work, the adjustment should be checked constantly. The first relation given above cannot be established directly but requires three adjustments.

First Adjustment—*Making the line of collimation parallel to the bottom element of the collars, or collimating the instrument.* Clamp the instrument, and unclip the collars. Sight at a distant point (as far as distinct), bringing the horizontal cross hair on it. Carefully turn the telescope in the collars by one-half a revolution around its axis, and sight again. If the horizontal cross hair is still on the sighted point, the telescope is collimated with regard to that cross hair; if it is off the point, bring it halfway back by means of the capstan-headed screws and the rest of the way by the plate screws. Repeat the operation over another point. Collimate it with regard to the other cross hair. Leave the screws at a snug bearing.

Second Adjustment—*Setting the bubble line in a plane with the bottom element of the collars.* Unclip the telescope, and clamp the instrument over a pair of plate screws. Center the bubble by means of the plate screws. Carefully and slowly turn the telescope in the collars in a small arc to the right, then to the left. If the bubble moves from center, bring it back by means of the azimuth or side screws.

Third Adjustment—*Setting the bubble line parallel to the bottom element of the collars.* Unclip the telescope, and clamp the instrument over a pair of plate screws. Center the bubble by means of the plate screws. Carefully take the telescope up, replacing it carefully in the Y's in the opposite direction, that is, the objective sighting in the direction where the eye piece originally was. If the bubble has moved, bring it back halfway by means of the altitude or foot screws of the spirit level and the rest of the way by the plate screws. Repeat in another direction until the adjustment is satisfactory.

The second relation is established by making the bubble line stay in the center of the graduation during a complete revolution of the instrument around its spindle.

Fourth Adjustment—*Making the axis of the instrument (not of the telescope) vertical.* Pin the clips; clamp and center the bubble over a pair of plate screws. Reverse the telescope over the same pair of plate screws; bring the bubble halfway back (if it has moved) by means of the plate screws.

Fifth Adjustment—*Making the bubble remain centered during a full revolution of the instrument.* Center the bubble, and revolve the instrument horizontally by a one-half revolution. If the bubble moves, correct it halfway by means of the support screws (at the foot of the Y's). If the rings become worn and unequal, use the two-peg method of the dumpy level.

The dumpy level, so called because of its compactness, is shown in Fig. 6–3. It is used mostly in England, although it is used to some extent in the United States because of the better stability of its adjustments over the wye level. The dumpy level differs from the wye level mainly in that the telescope of the dumpy level is permanently attached to the supports or uprights, but these uprights are adjustable. The two-peg adjustment method is as follows:

Drive two stakes (pegs) several hundred feet apart. Set the instrument approximately halfway between them. Level up and sight the rod, which is held in succession on each stake. The difference in the readings is the true difference of the elevation of the stakes, even if the instrument is not in proper adjustment. To test the instrument, set it near one of the stakes (the highest one, for instance); level up and sight the rod held on the other stake. Subtract the height of the instrument from the reading; the difference should be equal to the difference of elevation of the stakes as previously found. If these differences are not equal, set the target halfway between these readings, sight on it, and center the bubble by means of the altitude screws. Repeat the operation until satisfaction is obtained.

Centering the Objective and Ocular—These adjustments are made permanently by the manufacturer. Usually, four screws hold the tubes carrying the glasses; their heads pass through the outside tube where, after permanent adjustment, they are covered by a metallic ring.

Parallax—This is the apparent motion of the cross hairs on the object sighted when the eye is moved slightly. It shows the imperfect focusing of the ocular over the cross hairs. To correct this condition, hold a white surface, such as that of a piece of paper, slightly in front of the objective, and move the ocular tube in and out until the cross hairs are perfectly defined.

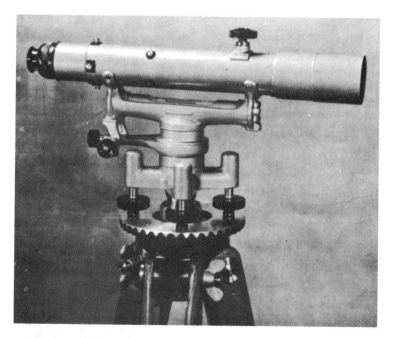

Fig. 6–3. A dumpy level.

Leveling Rod

This instrument, used in leveling, is usually 6½ feet high, graduated to hundredths of a foot and provided with a sliding target. The rod is made in two parts, arranged so that its length can be extended to 12 feet. Precision rods are of one-piece construction and have no target. Builders' rods may be graduated in feet, inches, and eighths of an inch, with a vernier reading in 64ths of an inch. A sliding disc, called a target, is provided with a vernier for extremely accurate work, reading to thousandths of a foot.

In use, the rod is held in a vertical position with its lower end resting on the point, the elevation of which is desired; the target is then moved up and down until its center coincides with the cross hairs in the telescope of the level. The reading of the elevation is made from the rod on a line corresponding with the center line of the target. There are various kinds of rods; some are designed to be read by the rodman, while others can be read through the telescope of the level.

Methods of Leveling

The simplest type of leveling is to find the difference in level between two points that are visible from a third point, the difference in level being less than the length of the leveling rod (Fig. 6–4). Set up and level the instrument at some point approximately halfway between the two points. Have the rodman hold the rod vertically on one of the points, and move the target up and down until its center coincides with the cross hairs of the level. Take a reading; this is the HI, or height of the instrument above the bench mark A. Turn the telescope on its spindle, have the rod held on the other point, and take a similar reading at B. The difference in level is equal to the difference in the readings.

If the difference in level is greater than the length of the rod, the method shown in Fig. 6–5 is used. Divide the distance between the two points into sections of such length that the difference in level between the dividing points A, B, and C, called stations, are less than the length of the rod. Set up and level between A and B, and measure the distance Aa, which is called backsight. Then, reverse the telescope, and take reading Bb, which is called foresight. Next, set up and level between B and C, and take readings Bb' and

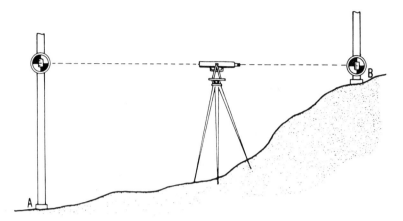

Fig. 6-4. Leveling between two points whose difference in level is less than the length of the rod.

Cc. Repeat the operation between C and D, taking readings Cc' and Dd. The difference in level between stations A and D is equal to the sum of the differences between the intermediate stations; that is, this difference equals $(Aa - Bb) + (Bb' - Cc) - (Cc' - Dd)$, or $(Aa + Bb' + Cc') - (Bb + Cc + Dd)$.

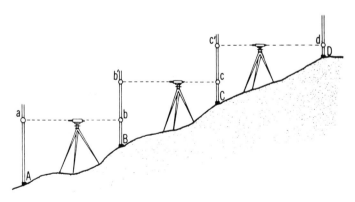

Fig. 6-5. Leveling between points whose difference in level is greater than the length of the rod.

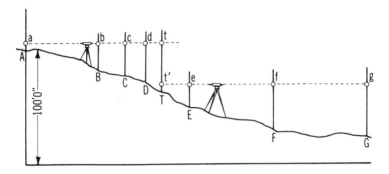

Fig. 6–6. Finding the relative elevations of several points in rough terrain.

Usually one wants to find the relative elevations of several points, as in grading work, in which case it is necessary to keep more elaborate notes and to measure distances between the stations. The method employed for this type of leveling is shown in Fig. 6–6, and the field notes are recorded as shown in Table 6–1.

Assume a datum or reference line below the elevation of the lowest station, and refer all elevations to this line. Start at some permanently fixed point, such as a mark on a building or the top of a hydrant; this is called a bench mark. Let A in Fig. 6–6 be the bench mark, and assume a datum line 100 feet below the level of A. Start with the instrument between A and B, and take a backsight on A. The distance Aa is found to be 4.2 feet, which, when added

Table 6–1. Field Notes
(Corresponding to the operations illustrated in Fig. 6–6)

Station	Distance	Backsight	Height of Instrument	Foresight	Elevation	Remarks
A	0	4.2	104.2	....	100.0	Bench
B	100	....	....	10.1	94.1	mark,
C	60	....	....	7.3	96.9	top of
D	50	....	....	5.8	98.4	hydrant.
T	....	4.1	99.1	9.2	95.1	Turning
E	70	6.8	....	....	92.3	point
F	110	....	....	9.5	89.6	
G	80	....	....	11.5	87.6	

to 100 feet, gives the height of the instrument. Next, take foresights on B, C, and D, and record these readings in the proper column. Readings Bb, Cc, etc., subtracted from the height of the instrument, will give the elevations at B, C, etc. This is done, and the results are recorded in the proper column of field notes. The ground falls away so rapidly beyond D that it is necessary to set up the level farther along and, therefore, establish a new height of instrument. This is done by holding the rod at some convenient point, such as at T, called the *turning point*, and taking a foresight, which measures the distance Tt (9.2 feet). The level is then set up in its second position between E and F, and a backsight is taken on the rod in the same position, which gives the distance Tt' (4.1 feet). The distance $t't$ then equals $9.2 - 4.1 = 5.1$ feet, and this is subtracted from the previous height of instrument, thus giving the new HI, which is $104.2 - 5.1 = 99.1$ feet. A backsight is now taken on E, and foresights are taken on F and G. These are recorded in the proper columns, and the elevations are found by subtracting these distances from the new HI. The horizontal distances between the stations are measured with a tape and recorded in the second column. When plotting a cross section from notes kept in this manner, the datum line is drawn first, and perpendiculars are erected at points corresponding to the different stations. The proper elevations are then indicated on these vertical lines, and a contour line is drawn through the points so marked.

Directions for Using Level

Note carefully the following mode of procedure in leveling:

1. Center the bubble over one pair of plate screws, then over the other pair. Plate screws should have a snug bearing. When looking at the bubble or at the cross hairs, the eyes should look naturally, that is, without strain. Try to observe with both eyes open.
2. Adjust the eye piece to the cross hairs for parallax.
3. Turn the instrument toward the target. It is better to level up facing the target.
4. Look again at the bubble.
5. Sight the target, and have it set right by motions according to a prearranged code with the rodman.

6. Look again at the bubble.
7. Read the rod or direct the target from the intersection of the cross hairs only.
8. Approve the target when absolutely sure.
9. Have the height of the target called out by the rodman.
10. Enter this height in the field book.
11. Quickly make needed calculations.
12. Briskly motion the rodman to a new station or to stay for a turning point and backsight, and move yourself to another position.

The following additional hints will also be found useful:

Guarding against the sun—Draw the telescope shade, or use an umbrella or a hat.

Length of sights—Avoid sights too short and too long; 250 to 350 feet should be the limit of the sights.

Equal sights—The length of the backsight should practically be the same as the length of the foresight; this may be approximated by pacing, or by sighting with the stadia cross hairs in the telescope.

Long sights—When sights longer than the maximum allowable in one direction only are unavoidable, correction should be made for curvature.

Leveling up or down a steep slope—The leveler, after some practice, will place his instrument so as to take a reading near the top or the bottom of the rod (as the case may be), thus gaining vertical distance, but this produces unequal sights. He may also follow a zig-zag course.

Leveling across a large body of water—*1. A running stream.* Drive a stake to the water surface on each side of the stream and in a direction normal to the flow, although the line may not run so. Take a foresight reading on the first, a backsight reading on the second, and continue to and along the line. The elevations of the two stakes may be assumed equal. *2. Across a pond.* If a pond or lake is too wide to ensure a good sighting across, use essentially the same method as for a stream. Drive stakes on each side and to the water surface; take a foresight reading on the first and a backsight reading on the second.

Across a wall—Take a foresight reading on the rod set on a stake, driven to the natural surface on the first side of the wall. Measure the height of the wall above the stake, and enter it as a backsight reading. Drive a stake to the natural surface on the second side of the wall, measure the height of the wall on that side above the stake, and enter it as a foresight reading. Set the rod on the stake, and take a sight on it, which will be a backsight reading. Continue using this method until the leveling has been completed.

In underbrush—If it cannot be cut down on the line of sight, find a high place or provide one by piling logs, rocks, etc., to set the instrument on.

Through swamp—Push the legs of the tripod down as far as possible. The leveler lies on his side. Two men may be necessary at the level. If the ground is still unsafe, drive stakes or piles to support the instrument.

Elevations to be taken at road crossings—Take elevations both ways for some distance.

Elevations to be taken at river crossings—Take elevations of high-water marks and flood marks, with the dates of same. Question residents for these dates and also for dates and data of extreme low water.

Proper length of sights—This will depend on the distance at which the rod appears distinct and on the precision required. Under ordinary conditions, sights should not exceed 300 feet where elevations are required to the nearest .01 foot, and even at a much shorter distance, the boiling of the air may prevent a precision reading of this degree.

Correction for refraction and earth curvature—A level line is a curved line at which every point is perpendicular to the direction of gravity, and the line of sight of a leveling instrument is tangent to this curve. This makes it necessary to take this curve into account in some leveling operations. If reasonable care is used to make the lengths of backsights and foresights approximately equal, this aberration is self-correcting, but in extremely long lines, it is approximately 2 inches in one-half mile, or about $\frac{2}{3}$ d^2, in feet, where d is equal to the distance, in miles. This correction is usually combined with that for refraction. The combined correction is 547 d^2, and it is *negative*.

Trigonometric Leveling

Finding the difference in elevation of two points by means of the horizontal distance between them and the vertical angle is called trigonometric leveling. It is used chiefly in determining the elevation of triangulation stations and in obtaining the elevation of a plane-table station from any visible triangulation point of known elevation. In triangulation work, the vertical angles are usually measured at the same time the horizontal angles are measured, so as to obtain the elevations of triangulation points as well as their horizontal positions. The vertical angle is measured to some definite point on the signal whose height above the center mark of the station was determined when the signal was erected; the height of the instrument above its station should be measured and recorded. In the most exact work, the angles are measured with a special vertical circle instrument. In less precise work, an ordinary Theodolite, whose vertical arc reads by verniers to 30 seconds or to 20 seconds, may be used, but with such instruments only single readings can be made. The best results with such an instrument are obtained by taking the average of several independent readings, one-half of which are taken with the telescope direct and the other half with the telescope inverted. In every case, the index correction, or reading of the vertical arc when the telescope is level, must be recorded.

The Transit

This instrument is designed and used for measuring both horizontal and vertical angles. It consists of a telescope mounted in standards which are attached to a horizontal plate, called the limb. Inside the limb, and concentric with it, is another plate, called the vernier plate. The lower plate or limb turns on a vertical spindle or axis which fits into a socket in the tripod head. By means of a clamp and a tangent screw, it may be fastened in any position and made to move slowly through a small arc. The circumference of this plate is usually graduated in divisions of either one-half or one-third of one degree, and in the common form of transit, these divisions are numbered from one point on the limb in both directions around to the opposite point, which is 180 degrees. The graduation is generally concealed beneath the plate above it, except at the verniers. This upper plate is the vernier plate, which turns on a spindle fitted into a socket in the lower plate. It is also provided with a clamp by means of which it can be held in any position, and with a tangent screw by which it can be turned through a small arc. A vernier is a device for reading smaller divisions on the scales than could otherwise be read.

The transit is generally provided with a compass so that the bearing of any given line with the magnetic meridian may be determined, if desired. It also has a spirit level attached to the telescope, so that it may be brought to a horizontal position and made to serve as a level. A typical transit is shown in Fig. 6–7.

Construction of the Transit

The general features of the transit construction are shown in Fig. 6–8; referring to the figure, these are briefly as follows:

Parallel Plates—There are two plates, one upper and one lower. The lower plate *A* is generally formed with two parts. The outside part is a flat ring and is screwed to the tripod head. The inside part is another flat ring of a diameter larger than the open-

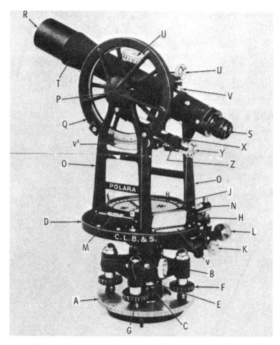

Fig. 6–8. The transit. In the illustration, A represents the lower plate; B, the upper plate; C, the central dome; D, the divided limb; E, the spindle; F, the foot screws; G, foot-screw cups; H, the vernier plate; J, the compass circle; K, the clamp-screw vernier plate to divided limb; L, the tangent screw; M and N, spirit levels; O, standards or supports; P, the horizontal shaft; Q, the vertical arc; R, the objective; S, the ocular; T, the telescope; U, racks and pinions; V, the adjustable cross-hair ring; v, the divided-limb vernier; v', the vertical-arc vernier; X, the spirit level; Y, the gradienter; and Z, the scaled index.

ing in the outside part and has a central dome *C*, which is perforated on the top. The inside part is movable and rests on the under side of the outside part. The upper plate *B* is generally made in the form of a central nut, with four arms at right angles (or three at 120°). The upper plate carries an inverted conical shell, the lower portion of which passes through the perforation in the dome of the inside part of the lower plate, where it expands into a spherical shape and thus forms a ball joint with the lower plate. This spherical member is perforated in the center to allow the passage of a plumb-bob string.

Foot Screws—The two plates are connected by four (sometimes only three) foot screws *F* in order (1) to clamp the lower and upper plates, making them fast with each other and with the inverted shell, and (2) to serve in leveling the instrument. The screws pass through the ends of the arms of the upper plate and are surmounted by dust caps. Their feet fit into small cups *G*, which rest on the top surface of the lower plate to avoid wear.

Shifting Center—Since these cups, as well as the central part of the lower plate, may be moved (after slightly loosening the foot screws), a slight motion may be given to the instrument to better set it over a given point of the ground. This arrangement is called a shifting center.

Outer Spindle—A second conical shell fits and may revolve in the conical shell attached to the upper plate. It is the outer spindle, and it carries projections to form attachments with the other parts of the transit.

Divided Limb—The upper portion of the outer spindle terminates in a horizontal disc of plate *D*, the limb of which is divided into 360°, subdivided into one-half, one-third, or one-quarter of one degree. Every ten degrees are numbered, either from 0° to 360° or from 0° to 180°, either way; the degree marks are a little longer than the subdivisions, and every fifth degree has a mark slightly longer yet.

Lower Motion—The outer spindle and the divided limb are also called the lower motion.

Inner Spindle—A solid inverted cone fits into the outer spindle and may revolve in it; it is the inner spindle, and, like the outer one, it is provided with some projections for similar purposes.

Vernier Plate—The upper portion of the inner spindle projects farther than the divided limb and also carries a horizontal disc H, which moves in a plane parallel to the divided limb (which it covers), except for two rectangular openings in opposite directions through which the divisions of the limb may be seen. These openings each carry a vernier v by means of which the subdivisions of degrees are again divided. Some verniers read to 1 minute, others to 0.5 minute, and some to 10 seconds. To facilitate the reading of the vernier, the openings are sometimes fitted with a reflector and a magnifying glass.

Upper Motion—The inner spindle and vernier plate H are also called the upper motion. The vernier plate carries a compass circle, shown at J.

Compass Circle—This consists of a circular box, the bottom of which carries at its center a sharp pivot of hard metal (hard steel or iridium) on which a magnetic needle approximately 5 inches long is balanced by an agate cup fixed in the middle of its length (Fig. 6–9). The needle is strongly magnetized; its north end is distinguished by color or ornamentation, and its balance is regulated by

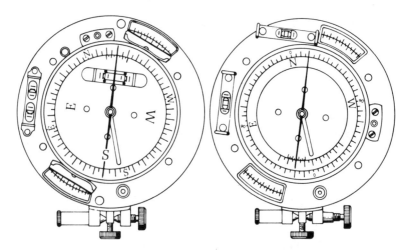

Fig. 6–9. Two typical compass boxes. The compass circles are graduated to one-half degrees and numbered in quadrants. The variation plate is provided so that the magnetic declination may be set accurately; the cardinal points shift with the graduated circle. *(Courtesy C. L. Berger & Sons, Inc.)*

a small coil of fine wire wound around one arm, which can be shifted. The limb, which is formed by the edge of the sides of the box, is divided into 360° with half degrees shown; they are numbered from two zeros marked at the ends of a diameter to 90° right and left. The bottom of the box is marked with two rectangular diameters corresponding to the graduations 0° and 90° of the vernier and two other diameters at 45° to the first. The forward end of the diameter marked 0° is designated by the letter N, and the rear end is designated by the letter S, corresponding to north and south. The ends of the transverse diameter marked 90° are designated by the letters E on the left and W on the right, corresponding to east and west. Note that this designation is the reverse of the standard mariner's compass. Since the telescope is fixed to sight from south to north, the compass indicates the direction of the sighting. When set to an ordinary surveyor's compass, the forward end of the frame carries a vernier and a tangent screw to read fractions smaller than ½ degree.

Controlling Clamps—A screw K permits clamping of the vernier plate H to the divided limb D. Another screw attached to the upper plate permits clamping of the divided limb to the upper plate.

Tangent Screws—One tangent, or slow-motion, screw L accompanies each clamp screw. It is used to complete the clamping at the exact spot where the clamp is to be made.

Spirit Levels—The spirit levels are attached to the vernier plate—one level M in front (north point of the box), the other N on the side—thus forming an angle of 90°.

Standards—The vernier plate carries two vertical standards or supports O, which are shaped like an inverted V and placed one on each side. The center of their legs is just opposite the 90° graduation of the compass box. They are made equal.

Horizontal Axis—The standards carry between and on the top of them a movable horizontal axis P.

Vertical Code—To the horizontal axis is attached, by means of a clamp screw, a vertical circle or arc Q, which is divided like the horizontal circle, and which in its vertical motion just touches a circular vernier v' carried by the left standard together with a slow-motion screw.

Telescope—In the middle of the horizontal axis and perpendicular to it is attached a telescope T of a description similar to that of the engineer's level, with an objective R and an ocular S, racks and pinions U for their motions, and an adjustable crosshairs ring V, with ordinary and stadia hairs.

Telescope Level—An adjustable spirit level X is also attached to the under part of the telescope, as in the engineer's level. This permits the transit to also be used as a leveling instrument, if necessary.

Motions of the Telescope—The telescope can function over the full range of the horizon and can measure any horizontal angle. Also, since the telescope is on a horizontal axis endowed with free motion, it may move in a vertical plane carrying with it the vertical arc, and it can therefore measure vertical angles. In the horizontal motion, the vertical cross hair of the telescope is brought exactly on the point sighted by means of the slow-motion screw L, carried by the vernier plate H. In the vertical motion, the horizontal cross hair of the telescope is brought exactly on the point sighted by means of the slow-motion screw carried on the inside of the left hand support and by moving the vertical circle.

Lines of a Transit

The following are the principal lines of a transit:

1. Vertical axis.
2. Horizontal axis.
3. Plate level line.
4. Attached level line.
5. Line of collimation.

Vertical Axis—The vertical line that passes through the center of the spindle E.

Horizontal Axis—The axis P of the shaft by which the telescope rests on the supports; it must be made horizontal.

Plate Level Line—The top or bottom lines of the plate level case N. These are level when the bubble is centered.

Attached Level Line—The level line of the bubble level X attached to the telescope; it is employed only when the instrument is used as an engineer's level.

Line of Collimation—The line determined by the optical center of the objective and the intersection of the cross hairs.

Relations Between the Lines of a Transit—The following relations must be obtained:

1. The plate levels must be perpendicular to the vertical axis.
2. The line of collimation must be perpendicular to the horizontal axis.
3. The horizontal axis must be perpendicular to the vertical axis.
4. The attached level line and the line of collimation must be parallel.
5. The zero of the vertical circle must correspond to the zero of the vernier when the telescope is horizontal.

Adjustments of the Transit

The following are the necessary adjustments of the transit.

First Adjustment—*Making the axis of the spindle vertical and the planes of the plates perpendicular to it.* Set one level over a pair of plate screws; the other level will thus be set over the other pair. Level up both levels by means of the plate screws. Turn the vernier plate around by a one-half revolution. If the bubbles remain centered during the motion, the vernier plate is in adjustment; if they have moved, bring them halfway back by means of the adjusting screws and the rest of the way by means of the foot screws. Repeat the operation, and determine if the bubbles remain centered when revolving the divided circle; if they do not, the plates are not parallel, and the transit must be sent to the manufacturer for repairs.

Second Adjustment—*Collimating the telescope.* Set up the transit in the center of open and practically level ground. Carefully level the instrument. Drive a stake or pin approximately 200 or 300 feet away; measure the distance. Take a sight on that point, and clamp the plates. Revolve the telescope vertically (in altitude) by one-half a revolution, thus reversing the line of sight. Measure in the new direction the same distance as first measured, and drive a pin. Unclamp and revolve the vernier plate by one-half a horizontal revolution. Sight again at the first point, and clamp. Again, revolve the telescope vertically by one-half a revolution. If the line of sight falls on the pin, the telescope is collimated; if not, drive a new pin on the last sight at the same distance as before, and drive another pin at one-fourth the distance between the first pin and the second. Move the vertical cross hair by means of the capstan-headed screw and an adjusting pin, until the intersection of the cross hairs covers the last pin set. Repeat the operation to be certain of collimation.

Third Adjustment—*Adjusting the horizontal axis so that the line of collimation will move in a vertical plane.* Level up carefully and sight on a high, well-defined point, such as a corner of a chimney, and clamp. Slowly move the telescope down until it sights the ground, and drive a pin there. Unclamp; revolve the vernier plate one-half a revolution, and revolve the telescope vertically one-half a revolution, thereby reversing the line of sight. Look again at the high point, and clamp. Slowly move the telescope down until it sights the ground. If the intersection of the cross hairs covers the pin, the horizontal axis is in adjustment; if not, correct halfway by means of a support-adjusting screw and the rest of the way by means of the plate screws. Repeat the operation, and verify the adjustment.

Fourth Adjustment—*Making the line of collimation horizontal when the bubble of the attached level is centered.* Drive two stakes 300 to 400 feet apart, and set up the instrument approximately halfway between these stakes. Level up and take readings on the rod held successively on the two stakes; the difference between the readings is the difference of elevation of the stakes. Next, set the transit over one of the stakes, level up and take a reading of the rod held on the other stake; measure the height of the instrument. The difference between this and the last rod reading should equal the difference of elevation as previously determined; if it does not, correct the error halfway by means of the attached level-adjusting screw. Repeat the operation, and verify the adjustment.

Fifth Adjustment—*Making the vernier of the vertical circle read zero when the bubble of the attached level is centered.* Level up the instrument. Sight on a well-defined point, and take note of the reading on the vertical circle. Turn the vernier plate one-half a revolution, and also turn the telescope vertically one-half a revolution; again, sight on the same point. Read and record the reading on the vertical circle. One-half the difference of the two readings is the index error, which may be corrected by moving either the vernier or the vertical circle, or the error may be noted and applied as a correction to all measurements of vertical angles.

It will sometimes be necessary to adjust the compass. When the adjustment is required, it may be accomplished by using the following procedure:

First Adjustment—*Straighten the needle.* Examine to see if the ends of the needle are set on opposite divisions; if not, fix the pivot so that they will. Revolve the box by one-half a revolution; if the needle does not set on opposite divisions, bend both ends by one-half the difference.

Second Adjustment—Place the pivot in the center of the plate. If the needle is straight, move the pivot until the needle sets on opposite divisions at points such as 0°, 45°, and 90°.

Instructions for Using the Transit

The transit requires various adjustments, as explained in the preceding section. To center the transit over a stake, rest one leg of the tripod on the ground, then grasp the other legs and place the instrument as nearly over the stake as possible. Then attach the plumb bob, and center it accurately by means of the shifting head. Avoid having the plates too much out of level, because this will result in unnecessary straining of the leveling screws and plates. Once the instrument has been centered over the stake, level it up by the spirit levels on the horizontal plate. To do this, turn the instrument on its vertical axis until the bubble tubes are parallel to a pair of diagonally opposite plate screws. Then, stand facing the instrument, and grasp the screws between the thumb and forefinger; turn the thumbscrew in the direction the bubble must move. When adjusting the screws, turn both thumbscrews in or out, never in the same direction. Adjusting one level will disturb the other, but each must be adjusted alternately until both bubbles remain constant.

The method of measuring a horizontal angle is shown in Fig. 6–10. The process of laying off a given angle is similar to that of measuring the angle. The transit is set up at the vertex of the angle, the vernier is clamped at zero, and the telescope is pointed at the target, thereby marking the direction of the fixed line. The limb is now clamped, the vernier is unclamped, and the vernier plate is

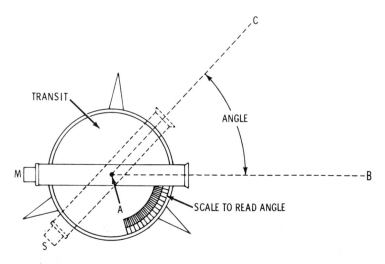

Fig. 6-10. The general principle of transit work. The transit is placed over the apex A of the angle CAB which is to be measured. The telescope is sighted to stake B (position M), and a reading is taken; it is then turned horizontally and sighted to stake C (position S), and another reading is taken. The difference between these readings gives angle CAB.

turned through the desired angle and clamped. A stake should now be driven in line with the vertical cross hair in the telescope, thus establishing the two sides of the angle. When laying out the foundations of buildings, a corner stake is first located by measurement, then the direction of one of the walls is laid out by driving a second stake. This direction may be determined by local conditions, such as the shape of the lot or the relation to other buildings. If the building is to be an extension to, or in line with, another building, the direction can be obtained by sighting along the building wall and driving two stakes in line with it. If it is to make a given angle with another building, this angle can be laid off as shown in Fig. 6–11.

Gradienter

Some transits carry a device called a gradienter (Fig. 6–8Y), which is attached to the horizontal axis by means of a clamp screw

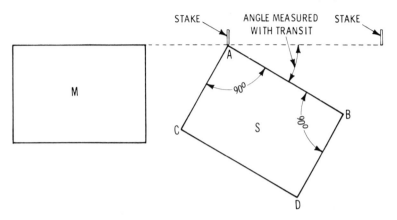

STAKE — ANGLE MEASURED WITH TRANSIT — STAKE

A

90°

90°

M

C

S

B

D

Fig. 6-11.   The method of laying out a new building (S) at a given angle with an old building (M). After the corner and the direction of one wall are determined, a right angle may be laid off (if the building is rectangular), thus locating two of the sides (AB and AC). The length of side AB is then measured, thereby locating corner B. The transit is set up at B, and line BD is laid off at right angles to AB. AC and BD are then laid off by the proper length, and the four corners of the building are thus located. If the building had not been rectangular, the proper angles could have been laid off instead of right angles.

and inside of the right-hand support. It is designed and employed for the determination of grades and distances, and consists of an arm in the shape of an inverted Y with curved branches. To the extremities of this arm are attached an encased spiral spring and a nut through which moves a micrometer screw with a graduated head that revolves in front of a scaled index (Fig. 6–8Z), which is also carried by the arm. The ends of the screw and the spring are on opposite sides of a shoulder that is carried by the right-hand support. The head is divided into tenths and hundredths, and every revolution moves it in front of the scale by one division, so that the scale gives the number of turns of the screw, and the graduated head gives the fraction of a turn.

In grading, if one revolution of the screw moves the cross hair a space of 1 foot on a rod held 100 feet away, the scope indicated by the telescope is 1 percent. To establish a grade, level up the telescope, clamp the arm of the gradienter, and turn the micrometer screw by as many divisions as are required in the grade. For instance, to set the gradienter at 2.35, move the head two complete turns plus 35 subdivisions. Measure the height of the telescope from the ground; set the rod at that height. Then hold the rod at any point on the line, raising it until the target is bisected by the cross hairs; the foot of the rod will then be on the grade.

Care of Instruments

With proper care, the usefulness of an instrument can be preserved for many years; therefore, the following suggestions on the care of instruments should be noted. The lenses of the telescope, particularly the object glass, should not be removed, since this will disturb the adjustment. If it is necessary to clean them, great care should be taken, and only soft, clean linen should be used. To retain the sensitivity of the compass needle, the delicate point on which it swings must be carefully guarded, and the instrument should not be carried without the needle being locked. When the needle is lowered, it should be brought gently on the center pin. The object slide seldom needs to be removed; when removal is necessary, the slide should be carefully protected from dust. Do not grease or oil the slide too freely; only a thin lubricant film is necessary. Any surplus of oil should be removed with a clean wiper. The centers, subject to considerable wear, require more frequent lubrication. After a thorough cleaning, they should be carefully oiled with a fine watch oil. All of the adjusting screws should be brought to a fine bearing, but they should never be tightened to such a degree that a strain is applied to the different parts; if this is done, the adjustment will be unreliable. When the instrument is carried on the tripod, all clamps should be tightened to prevent unnecessary wear on the centers.

The Stadia

This is a device that is used for measuring distances, and it consists essentially of two extra parallel hairs in addition to the ordinary cross hairs of the transit or a level telescope (Fig. 6–12). The

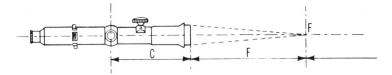

Fig. 6-12. The principle of stadia operation. The fixed stadia hairs are set so that they will intercept 1 foot on a rod at a distance of 100 feet. Since the image of the cross hairs is projected to a point beyond the telescope objective equal to its focal length, the rays of light converge at that point, and measurements must begin from there. Therefore, a constant must be added to all stadia readings equal to the focal length of the object lens plus the distance from the face of the objective to the center of the instrument. This constant is the factor F + C; for transit telescopes, it is equal to approximately 1 foot.

stadia hairs may be adjustable, or they may be fixed permanently on the diaphragm.

When using the stadia, distances are measured by observing through the telescope of a transit the space, on a graduated rod, included between two horizontal hairs, called stadia hairs. If the rod is held at different distances from the instrument, different intervals on the rod are included between the stadia hairs. The spaces on the rod are proportional to the distances from the instrument to the rod so that the intercepted space is a measure of the distance to the rod. This method of measurement furnishes a rapid means of measuring distances when filling in details of topographic and hydrographic surveys.

Most transits, all plane-table alidades, and some precision leveling instruments are fitted with stadia hairs. Stadia surveying has the advantage in that the intervening country does not have to be taped, and it provides a means of measuring inaccessible distances, such as across water and up steep hills and bluffs. If is well adapted to preliminary surveys for highways and railroads because the errors tend to be compensating rather than cumulative, but it should not be used for short distances, such as farms and city lots. In sights of 200 to 400 feet, it is possible to read a rod to the nearest hundredth of a foot, which represents one foot in distance. At 600 to 1200 feet, it is possible to read to the nearest hundredth of a yard, which represents 3 feet in distance. This is the precision to be expected in stadia measurements. The rod used is preferably a one-piece stadia rod, (Fig. 6–13F), but any standard leveling rod, except builders rods graduated in inches and sixteenths, may be used.

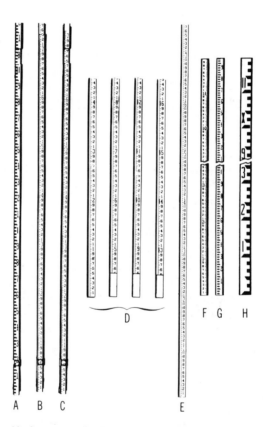

Fig. 6–13. Various popular leveling rods. In the illustration, A is the Philadelphia rod, in English graduations; B, the California rod; C, the Philadelphia rod, in metric graduations; D, the Chicago rod; E, the architect's rod; F, the stadia rod, in English graduations; G, the stadia rod, in metric graduations; and H, the broad stadia rod. *(Courtesy Eugene Dietzgen Co.)*

When leveling with an instrument which is equipped with stadia hairs, care should be taken not to confuse the center leveling cross hair with either of the two outside stadia hairs. It has been done. Although it is by no means obsolete, stadia surveying has now to a great degree been superseded by aerial photography. Neither is a substitute for careful taping.

For the vast majority of surveying purposes, the transit and levels described are more than sufficient. Three other devices can be even more useful, however.

The Theodolite (Fig. 6–14) is a transit that is more expensive but more accurate than a standard transit, and the Theodolite has more capacity. If you are doing readings over 500 feet, you may want to look into this.

The automatic level (Fig. 6–15) is also a tool used where great

Fig. 6-14.   Wild T-16 Theodolite. (Courtesy of Wild-Heerbrugg)

stadia, distances are measured by observing through the telescope of a transit the space, on a graduated rod, between the two horizontal hairs, called stadia hairs.

Review Questions

1. What are the three lines of the level?
2. What is a transit?
3. What are the fine lines of a transit?
4. Name the various leveling rods used for measuring distance.
5. Explain the terms "backsight" and "foresight" when used in leveling.

CHAPTER 7

The Design Process

The first consultation the designer has with the owner will usually result in the owner's setting forth what he or she wants or requires, such as the number of rooms, their sizes, the style of house preferred, the kind of materials he or she wants to use, and probably that all-important factor—how much money he or she will need, or how much is available. This first, or the first few, consultations may result in the designer's making a dozen or more sketches. It is the designer's function to guide the client in such matters, for often, indeed usually, the owner has no skill or aptitude for them. Later consultations may settle such things as lighting, heating, perhaps the quality and style of plumbing fixtures, and the make or quality of such appurtenances as air-conditioning, laundry, and dishwashing equipment. The designer usually makes sketches, probably freehands, for the owner's approval. Among the many particulars that should be settled before the final design work:

1. The owner may greatly prefer a house with a basement, but, perhaps for reasons of economy, he may be willing to accept a house on a concrete slab. The designer should be able to give advice as to the advantages and disadvantages of the two systems along with the comparative cost of construction.
2. The owner's mind may be set on a heating plant in the base-

ment, but in the event that a house with no basement is agreed on, he or she should understand that the heating plant must be in a utility room.

3. The owner may insist that there be a well-equipped laundry in the basement. The person who does the laundry may have some ideas on this subject.

4. The owner will probably insist that there be adequate closets for each bedroom. The owner may have some ideas as to exactly what "adequate" consists of.

5. The owner will want the kitchen to be convenient, with or without a garbage-disposal unit, probably with a dishwasher in a convenient location, possibly with room for a home freezer. Although "saving steps" is important, sufficient room in the kitchen may be more important. Few cooks appreciate a kitchen that is too small—it greatly inhibits style.

6. The owner may wish a dining room, and not just "dining space" in one end of the living room, and may think that a counter or bar and no partition between the kitchen and dining room is not sufficiently odor-resisting when cooking fish or corned beef and cabbage.

7. The owner, or perhaps the designer, may have some ideas concerning privacy in the home. There may be some objections to "window walls" that are supposed to "bring the outdoors indoors," and a preference may be expressed for baseball-proof walls instead; perhaps the idea of keeping large areas of glass clean is not appealing, and the owner would be quite content to leave the "outdoors" outdoors if there is plenty of living space indoors.

8. Then there is the constant problem of sound resistance. Modern homes are often noisy, with air-conditioning, forced-heating, laundry, and dishwashing equipment, attic fans, kitchen fans, bathroom fans, radio and television, and many other noise-generating sources. It is the designer's duty and obligation to see that such noises are isolated insofar as is possible. Partitions should be noise-resistant.

These are only a few of the problems of the modern home that the designer must understand; they are all with us and will probably be with us for many years to come. The owner probably does not, possibly cannot, understand how to handle such problems. The designer can, and should. It is the designer's function to guide the owner's ideas or simple notions so that the home environment will be satisfactory as far as the owner's means will permit. The designer will probably be blamed for any serious discrepancy, no matter if the owner *did* insist on it. "He shouldn't have allowed me to do *that*," etc.

It is not only desirable, but absolutely necessary, that the designer be familiar with the dimensions of the equipment, furniture, and other appurtenances found in the home. While it is rarely necessary that they be accurately detailed, space must be allotted for each one of them, and space is expensive.

Almost all modern building is governed by codes of some sort. In all government-financed homes, the government's minimum standards are strictly enforced, and city codes are often much more restrictive. Electrical codes occasionally seem to be unreasonable, but the building designer must be governed by them. Plumbing is often, *too* often, seriously skimped when no one is watching. Many states have plumbing codes, but they do not have the force of law unless augmented and enforced by local authority. There is a National Plumbing Code and while it is advisory, it is in line with good practice where there are no local codes. State and local boards of health may make it quite difficult for the designer of an inadequate plumbing system if the occasion arises.

For economy, kitchens and bathrooms should be placed back to back. A 3-inch copper soil pipe will fit into a partition of 2 × 4's whereas a 4-inch cast-iron soil pipe won't.

In perhaps most cases, the designer's duty is done when he prepares and delivers the drawings for a job. In some cases, he contracts to inspect the work at stated intervals, to assure that the work is satisfactorily done.

The designer should allow for the following thicknesses of walls in drawings:

1. Standard wood outside wall, ¾-inch plywood or insulating board sheathing, 3½-inch studs, ½-inch sheetrock inside, 4½ inches under the siding.
2. Inside partitions, 3½-inch studs, sheetrock ½ inch (both sides), 4½ inches.

3. Sound-resistant staggered-stud partitions, $3\frac{1}{2}$-inch studs staggered 2 inches, gypsum lath and plaster (both sides) $7\frac{1}{4}$ inches.

4. Single-width brick veneer, $\frac{3}{4}$-inch sheathing, $3\frac{1}{2}$-inch studs, $\frac{1}{2}$-inch sheetrock, $9\frac{1}{2}$ inches.

5. Concrete block, 8-inch plastered against the masonry, $8\frac{1}{2}$ inches.

6. Concrete blocks, 8-inch, $\frac{3}{4}$-inch furring, $\frac{1}{2}$-inch sheetrock, $9\frac{1}{4}$ inches.

7. Cavity masonry wall, $3\frac{3}{4}$-inch brick, $2\frac{1}{2}$-inch air space, 4-inch concrete blocks, plaster on masonry, $10\frac{7}{8}$ inches.

8. Chimneys—minimum wall thickness, $3\frac{3}{4}$-inch brick; liners, outside dimensions, $8\frac{1}{2}'' \times 8\frac{1}{2}''$, $8\frac{1}{2}'' \times 13''$, $8\frac{1}{2}'' \times 18''$, $13'' \times 13''$.

9. Ceiling heights—first floor, clear minimum, 7 feet 6 inches; basement, 6 feet 9 inches clear.

10. Stair wells—3 feet 2 inches × 9 feet, or clear head room 7 feet above nosing of treads, vertically.

An Example of Design

As an example of design, the series of illustrations presented in this section show the development of architectural drawings beginning with stock plans or plans that appear in newspapers and magazines from time to time. They give a prospective owner a good starting point, saving much time and study.

Certain things, however, should be kept in mind. The magazine drawings were doubtlessly prepared by a registered architect. While the publishing of a design may imply that he has given approval that the design can be copied, it is best to get the written consent of the original designer, or at least to secure this consent from the publisher. A registered architect's plans are protected from being copied, and court decisions have ruled that minor changes, regardless of how many of them there are, do not release the copier from liability. Purchasing a set of stock plans includes permission to build from the plans at least once. You might have to buy a second set to build the same design again. Determine where you stand, legally, before copying designs of any sort. It may save you from a costly and embarrassing situation.

Let us assume that we start with magazine plans (Figs. 7-1, 7-2). These two plans were prepared by a capable and experienced architect who has given us a practical and logical arrangement of a house plan, with stairs, doors, windows, closets, etc.; he has also indicated a proper and reasonable size for the various rooms. Any proposed changes or additions that the owner may desire may be taken up with the architect or builder and may be easily whipped into form on his drawing board.

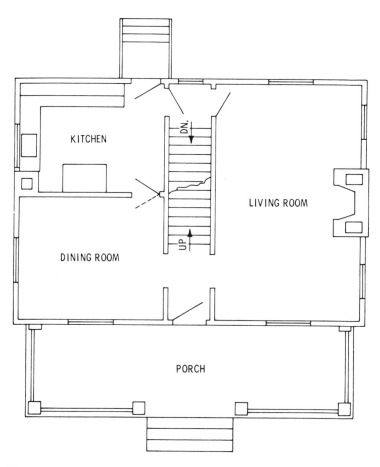

Fig. 7-1. First-floor plan from a magazine.

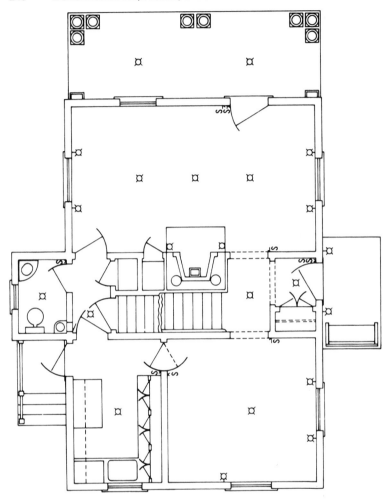

Fig. 7–3. The modified first-floor plan.

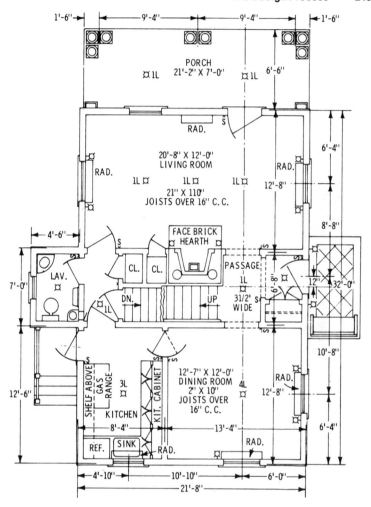

Fig. 7-4. The completed drawing of the first floor with all dimensions and specifications in detail.

signer must understand. The owner probably does not know how to understand and handle such problems. The designer should know how to handle these problems and guide the owner so that the home environment will be more than satisfactory when the work is completed.

It is absolutely necessary that a designer be familiar with the dimensions of the equipment, furniture, and other appurtenances found in the home. Electrical codes occasionally seem to be unreasonable, but the building designer must follow them. Plumbing is often seriously skimped unless local authorities have some kind of a ruling enforcing certain regulations.

Many times the future owner prefers to deal only with a carpenter or builder, and asks him/her to prepare the drawings and give an estimate on construction. The builder then calls the designer in for advice on the preparation of the drawing.

Review Questions

1. What must a designer know from his client before drawing up plans for a house?
2. Is it generally a good idea to change floor plans after construction has started?
3. Is the kind of material used in the construction important to the designer?
4. Is wall thickness, plumbing, and electrical wiring considered when drawing up plans?

CHAPTER 8

Specifications

By definition, a specification is a specific and complete statement detailing the nature and construction of the item to which it relates; as applied to the building trades, specifications describe briefly, yet exactly, each item in a list of the features and materials required to complete a contract for building an entire project.

Great care should be used when reading specifications to avoid misunderstandings and disputes. Each item entering into the construction is defined and described with precision so that there can be no chance of misunderstanding or double interpretation.

Example of Specifications

Specifications refer to the contract form of which they are a part. This saves repetition of statements with regard to liability of contractor, owner, etc. The following is an example of specifications.

Specifications

for

a frame dwelling

to be built for Mr./Ms. _____ of _____
in the county of _____ and state of _____
on lot number _____ on the _____ side of _____
street in the city or borough of _____, county of,
_____, state of _____. These
specifications relate to a set of attached drawings consisting of:

1. Plan of foundation and basement.
2. Plan of first floor.
3. Plan of second floor.
4. Plan of roof.
5. Four elevations.
   (eight sheets in all)

Detailed working drawings to be furnished as required, all of which are a part herewith, and are to be considered as such with any contract that may be made.

Height of Ceilings—The following dimensions for these heights are:

Basement to be................7 feet 2 inches clear of finish.
First story to be...............8 feet 6 inches between timbers.
Second story to be............7 feet 6 inches between timbers.

Interpretation of Drawings—For arrangement of floors, general finish, and measurements, reference must be made to the drawings. However, should any difference appear between the scale measurements and the figures, or between the wording of the specifications and the lettering on the drawings, the specification shall in all cases take precedence. If any error that is not explained either by reference to the drawings or specifications becomes apparent, the contractor shall refer them to the architect for correction before proceeding with the work.

Conditions—The contractor must see that all the work on the said building is performed in a thorough, workmanlike, and substantial manner by competent workmen and must also furnish all materials (the best of their respective kinds), labor, implements, transportation, etc., if not otherwise specified.

All painted parts of the exterior must have a prime coat of paint as fast as it is ready. The succeeding coats must not be applied within 3 days of the former, and then not in wet or freezing weather or other conditions specified by the paint manufacturer.

The contractor must protect all work while the building is in his hands, remove all superfluous materials or rubbish, and not obstruct the grounds around the foundation for grading and filling in as soon as the building is up. Figures are to take precedence over scale measurements.

Mason's Work—Excavate to the length, breadth, and depth required for the foundations, as shown on the architect's drawings. The top soil is to be removed and placed in a separate pile from the other excavated materials—25 feet away from the excavation where directed. Also, excavate for a septic tank and overflow 75 feet from the foundation, as will be directed, containing 28 cubic yards to be built of concrete, with baffles, drains, etc., all of which is set out in a special plan for same. The septic tank may be included in the plumbing subcontract. The main tank must be waterproof, although the overflow need not be. The tank shall be connected to the house at a point below the lowest fixture and below the frost line with a uniform declination of not less than 12 to 20 inches and will have no running U trap. The drain is to be made of 6-inch socket-jointed Transite tile laid in tight cement joints from a point 4 feet outside the foundation.

Foundation—Foundations and footings, as shown on the plans, are to be made of poured concrete with 8-inch walls and 16-inch footings.

Chimneys—Build two chimneys, as shown, of the same size and shape. Use an approved hard-red-pressed brick for all exposed parts of the outside chimney and for topping out. The fireplaces in the parlor, dining room, and bedroom are to be faced with the same brick (smooth inside the fireplaces), with 8 inches on the sides and 24 inches at the top. Buff the joints, and straighten the arches on $3\frac{1}{2}'' \times 3\frac{1}{2}''$ angle irons, unexposed. Use fire brick for the backs; lay the bricks in an approved refractory mortar (no fire clay). Spring trimmer arches for the hearths are to be laid with the same selected brick. All flues are to have tile linings, approved chimney pots, and clean outs.

Mortar—All mortar for brick work is to be grade specified by building code regulations.

Installing Drywall—All walls, partitions, and ceilings, and all studded and furred places in all stories, are to be covered with ½-inch-thick Sheetrock, which will have joints covered with tape and joint compound. Two coats of compound will be used where wallpaper is used, three coats where paint is to be used. Panels used shall be 4 feet by 8 feet and will be installed horizontally on walls, across the framing members on ceilings. The panels are to be secured with 1⅜-inch blued ring shank nails and pieces staggered. Inside corners of the paneling shall be covered with inside corner molding, with nails placed 7 inches apart. Outside corners shall also be covered with metal molding. All joints will be smooth to the touch.

Tiling—The floors of bathrooms will be tiled with $3'' \times 3''$ octagonal and 1-inch-square vitrified tiling, colors to be selected. The side walls will be tiled 4 feet high of plain white glazed $2\frac{1}{2}'' \times 4''$ molded base and nosing, with a narrow tinted stripe at the top of the sanitary base and under the nosing. The floors will be properly prepared by the carpenter by setting the rough floor ½ inch below the top of the floor beams. All tiling will be set in adhesive recommended by the manufacturer, and the floors will be finished flush with the wood finish floors.

Other Floors—There will be a concrete floor in the furnace room, in the shop, and in the area from the west end turning east to the cross wall, as shown in the plans. All floors will be 3⅝ inches minimum, with $6'' \times 6'' \times$ No. 11 reinforcing mesh. The kitchen hearth will be built in the same manner.

Coping—There will be 4-inch caps of blue stone on all piers showing them, edged on four sides, 3 inches larger than the piers. Cope area walls, which are to be 8 inches, with $2'' \times 10''$ blue stone where circular, fitted to radius. *No* patching of stone will be permitted.

Timber—All timber will be thoroughly seasoned, No. 1 common pine, square, straight, and free from any imperfection that will impair its durability or strength. *No* individual piece is to have a moisture content of more than 19%; the architect will check this.

Framing—The framing will be as indicated on the drawings. Headers over openings will be the sizes indicated in detail. *No header with a checked moisture content of over 15% will be acceptable.* Frame so that sheathing will be flush with the foundation wall. All moldings are to be miter-spliced and mitered at angles. *No* butt ends will be showing in the finish.

Timber sizes will be as follows:

Sills.................... 2″ × 8″
Girders................ 10″ standard I beams (25.4 pounds)
Corners............... 4″ × 6″ backed with 2″ × 4″ or built up
Main plate........... 4″ × 4″ (2″ × 4″ doubled)
Rafter plate.......... 4″ × 4″ (2″ × 4″ doubled)
Studding (general).. 2″ × 4″
Closet studding...... 2″ × 3″
Main rafters.......... 2″ × 6″
Dormer rafters....... 2″ × 4″
Ridge boards......... 1¼″ × 8″
First floor joists...... 2″ × 10″
Second floor joists... 2″ × 8″
Second story ceiling
  beams.............. 2″ × 6″

Spacing and Bridging—Place all studding, floor, and ceiling joists on 16-inch centers. In every span of flooring exceeding 10 feet, there will be a row of 1″ × 2″ bridging or 2″ × 3″ double nailed at each end. Rafters will be placed on 24-inch centers.

Partitions—All partitions are to be set plumb, well braced, and nailed; studs at all angles and openings are to be doubled, and extra block is to be set at door openings for base nailing. All partitions that are not supported below are to be firmly trussed and braced. Ceilings to all closets will be furred down to within 12 inches of the door head except in closets over 2 feet deep. There will be trued ⅞-inch grounds at top of base and around all openings.

Lumber—All outside finish lumber will be clear white pine unless otherwise specified. All exterior finish lumber is to be free from large or loose knots and will also be clear and thoroughly dry.

Sheathing and Sheathing Paper—Cover all the exterior walls with ¾-inch plywood sheathing nailed to each stud with 8d nails. With joints cut on studs or backed for end nailing, cover with

TYVEK®, well lapped and extending under all trim and around all corners to make a complete and tight job.

Exterior Finish—Windows, door casings, cornices, corner boards, water table, brackets, band courses, etc., are to be made to the detail furnished in the drawings. The stock moldings that are to be used are numbered on the drawings. The first story is to be covered with the best grade cedar lap bevel siding, laid at $4\frac{1}{2}$ inches to the weather. The second story and gables are to be covered with 18-inch hand-split and resawn shakes, laid at $8\frac{1}{2}$ inches to the weather. Use hot-dipped galvanized nails, whose length will be approved by the architect. Window casings will be laid $2\frac{1}{2}$ inches to the weather, and the front door frames and casement windows will be according to detail shown in the plan.

Shingling—Cover all roofs with $8'' \times 16''$ Pennsylvania blue slate, laid 7 inches to the weather. All hips and other parts that require it are to be made secure against leaks by the proper use of slaters' cement and proper flashings. An ornamental galvanized-iron ridge crest will be placed on the main ridge; see details on drawings for this crest.

Flashing—Flash around chimneys, over all doors and windows, heads exposed to the weather, and where roofs join walls with 16-ounce sheet copper. Do the same in all valleys and wherever required to secure a tight job. Each side of the valleys is to have a water check turned up 1 inch in the metal.

Flooring—First and second stories are to have double floors, with a subfloor of $\frac{5}{8}''$ CDX plywood. The first- and second-story finish flooring is to be oak tongue-and-groove strip flooring, which is to be thoroughly seasoned and blind nailed over building paper. There will be no joints in the main hall and only one joint in the run of boards in other rooms of the first floor. The second-story floors are to be cleaned and sandpapered to a smooth finish for the painter. Oak thresholds are to be set to all outside doorways, and hard rubber-tip door stops are to be located behind all doors that open against a wall.

Window Frames—These are to be made of seasoned white pine.

Sash—All sash and frames are to be made by the Johnson Corporation and are to be of kiln-dried, vinyl-sheathed white pine; the numbers are given on the drawings.

Screens—All windows that open are to be fitted with bronze- or copper-wire window screens.

Glazing—All sash and outside doors, where indicated, are to be glazed with Johnson insulated windows or their equivalent. All hall doors are to be glazed with French plate; the plate in the Dutch door will be beveled. The basement sash is to be glazed with a single strength glass.

Blinds—All windows, where indicated, are to be provided with an approved type of blind that will be $1\frac{1}{8}$ inches thick and made of the best grade of seasoned white pine; all blinds will move freely after painting. The blinds are to be hung on approved cast-iron blind hangers.

Door Frames—All inside door frames in finished parts of the house, first and second stories, are to be made of white pine $\frac{25}{32}$ inch thick, set plumb and true, and blocked in four places on each side. Outside door frames are to be rabbeted for doors. All frames are to be flush with the plaster finish.

Doors—Unless otherwise specified, all inside doors are to be made of slab-type birch veneer with hollow cores and will be $1\frac{3}{8}$ inches thick. Outside doors are to be $1\frac{3}{4}$ inches thick and will be made of solid-core, slab-type birch veneer. The front doors are to be of the design shown in the drawings. Hang all doors throughout with loose-joint ball-tip butts of sufficient size to throw them clear of the architraves. Doors are to have three $3\frac{1}{2}'' \times 3\frac{1}{2}''$ butts on $1\frac{3}{8}$-inch doors and three $4'' \times 4''$ butts on $1\frac{3}{4}$-inch doors. A hardware schedule will be furnished. Hang both double-swing butlery doors on double-acting brass spring hinges. Furnish all nails, except those used for inside work, galvanized and all other hardware that will be necessary for the completion of the work in the proper manner.

Interior Trim—For the basement, the interior trim is to be selected cypress or redwood. For the first and second floors, the trim is to be unselected birch. There will be a 4-foot 6-inch paneled wainscoting in the dining room, first floor hall, and up stairway; this panel will be made of $\frac{1}{4}$-inch birch plywood, with trim as shown in the details. There will be a 5-inch cabinet plate shelf in the dining room, the bottom member of which will be a picture molding; this shelf will match the door head trim.

Stairs—The main staircase is to be made of unselected birch.

The stringers and treads are to be $1\frac{1}{8}$ inches thick, as shown in details. The risers are to be $\frac{3}{4}$ inch thick. The risers and treads are to be housed into the wall stringer and return-nosed over the outside string. The rails are to be $3'' \times 3''$ molded, with ramps as shown in the details. Balusters will be $1\frac{5}{8}$ inches, taper turned, three to a thread, and proportionately more for increased widths. Newels and column newels are to be as shown in the details. The run on the first flight is to be $10\frac{1}{4}$ inches from face to face of the rise with 12-inch treads; the second flight and basement stairs are to be $9\frac{3}{4}$ inches of run with $11\frac{1}{2}$-inch treads. For the basement stairs, cut $2''$ $\times 12''$'s for the stringers and $2'' \times 10''$ yellow pine for the treads.

Mantels—There will be two mantels where indicated on the drawings. The contractor will figure them to cost $2000 each complete, including linings and face and hearth tile. This amount will be allowed the owner to use at his, her, or their option in the selection of same. The entire cost is to be figured in the contract price, including the setting of the mantels by the contractor.

Pantry Cabinets—There will be a cabinet, where indicated, with three glass doors, above the draining board; this cabinet will be 10 inches deep inside and will contain three shelves. The wall cabinets are to be constructed as shown in the details.

Closet Shelving—The trim on the inside of the cabinets is to be plain. There is to be an average of 10 feet of 12-inch shelving to a closet, with 6-inch clothes strips and 1 dozen clothes hooks, japanned. The kitchen closet and the closet under the kitchen stairs are to have suitable shelving and sufficient pot hooks and other fixtures. There will be 25 feet of shelving in the shop closet. There will also be 1 dozen clothes hooks under the basement front stairs.

Plumbing—All necessary materials for completing the plumbing installation, as hereafter set forth, in a correct and sanitary manner are to be included in the general contact. The state plumbing code shall be strictly followed.

Electric Wiring—No. 12 Romex sheathed cable is to be installed under and subject to the requirements and regulations of the *National Electrical Code* and all state, county, and municipal codes. The locations for all electrical outlets will be shown in the drawings.

Water Pipes—Water is to be brought from the street main into the house through $\frac{3}{4}$-inch copper tubing, or plastic if preferred.

Copper water tubing is to be used on all straight-line work. Place a hose-bib cock on the main at a point against the house for hose purposes, both front and rear, with a stop and waste cock in the basement. Complete all necessary digging for the laying of sewer and water pipes to the house; no trenches are to be less than 36 inches below the grade at any point. The pipes are to enter the house in the basement at a suitable point for intersection with the inside piping system. The dirt is to be well rammed over the pipe in the trenches as it is refilled. The house sewer is to be installed by the plumber and it will consist of a perfect 4-inch glazed socket-jointed tile pipe to a point exactly 4 feet outside the foundation wall; the soil pipe is to be 4-inch cast iron. All water pipes must have a gradual fall from the fixtures which they supply, and they must open at their lowest point for drainage purposes. All the cast-iron waste pipe will be furnished with the necessary fittings. Furnish and install one 60-gallon gas-fired water heater of an approved type and manufacture; this water heater is to be supplied with water through a ¾-inch copper tube. Place a shutoff cock on the supply pipe. Take hot water from the water heater to and over the kitchen sink and to all other fixtures, except toilets, with ½-inch copper tubing. The supply to the toilets will be through ⅜-inch copper tubing. Cold water to all fixtures will be through separate pipe lines. There must be no depressions in any pipe, and hot water must be kept rising from the boiler head.

Kitchen Sink—Furnish and install a cast-iron enameled kitchen sink, with a garbage disposer of (*specify make and model*). Furnish and install an automatic dishwasher (*specify make and model*) where indicated in drawings.

Wash Trays—Provide and install where indicated on drawings, one two-part stone tub on galvanized iron legs. Supply the tub with hot and cold water through ½-inch copper tube and brass faucets, one for cold water, threaded 1½ inches waste, with traps, plugs, and chains complete. The waste drain is to be connected with the soil pipe through a 2-inch copper pipe.

Bathroom Fixtures—All bathroom fixtures are to be (*specify name of manufacturer*) make, as listed in their catalog. The basement bathroom water closet will be (*name and model*); lavoratory, (*name and model*); bathtub, 5-feet-6-inch (*name and model*). The other toilets are to be (*name and model*). The two second-story baths are to be (*name and model*), 18″ × 27″. Before any wall finishing is done, all supply pipes must be proven perfectly tight by a satisfactory test, and they must be left perfect at the completion of the test.

Painting—The entire exterior woodwork, except shingles, is to be painted with three coats of (*manufacturer's name*) best grade of ready-mixed paints, thinned as necessary and as specified by the manufacturer. The color of paint will be plain white, except for the shutters and shakes, which will be green and will be given two coats of (*manufacturer's name*) shake and shingle finish. The second-story floors will be sanded smooth and given three coats of (*manufacturer's name*) polyurethane varnish. The first-story floors are to be left bare for wall-to-wall carpeting or vinyl tiling. All interior birch trim is to be finished natural, with one coat of white shellac and two coats of (*manufacturer's name*) of flat varnish. The type and color of the interior wall paint for all interior walls will be specified by the owner.

Condition of Bids—The owner reserves the right to accept or reject any or all bids.

Summary

A specification is a statement containing a detailed description or enumeration of particulars, as of the terms of a contract, and details of construction usually not shown in an architectural drawing. Great care should be used when reading specifications to avoid misunderstandings and disputes. Each item entering into the construction is defined and described with such precision that there can be no chance of misunderstanding or double interpretation.

Framing will be indicated on the drawings. Specifications as to type of timber used, such as No. 1 common yellow pine, square, straight, and free from any imperfections that will impair its longevity, will be indicated. Headers over window and door openings will be indicated as for size and installation procedure.

Interior trim, including stair casing, will be shown. Moldings used throughout the house will indicate miter angles and type of lumber used. Ceiling height, as well as plastered or dry-wall construction, will be specified. Most plastered walls will be specified as to type of finish coat, which in most cases is at the option of the owner, and will also specify that exposed corners be protected with metal corner beads.

All necessary materials for completing the plumbing installation, as set forth in the specifications, shall meet all state and local regulations. All electrical wiring will be listed and must meet the requirements and regulations of the *National Electrical Code* or state or local codes.

Review Questions

1. Why should great care be used when reading specifications?
2. What information is included in the specifications on framing a house?
3. What code should be followed when plumbing a house?
4. Can specification on a dwelling be changed or altered? If so, how?
5. What code should be followed when electric wiring is installed?

CHAPTER 9

Architectural Drawings

Drawings provide designers with a practical method of communicating their ideas to the carpenters and builders. Drawings are a type of shorthand that not only reflects, but also fixes on paper, the ideas of the designers. They help avoid possible misinterpretation.

Use of Drawings

Contractors and estimators should retain prints on jobs that they complete in order that they might be used to compare future jobs. They should also keep at least two sets of prints on a job in progress. One set is the working drawings for the tradespeople. On the second set, the lead carpenter adds the changes made as the structure was built, to be turned over to the owner to update the original drawings issued. The final drawings will identify any changes that were made during construction. The owner may change the original drawings to reflect the changes, so that the drawings will be fully up to date to assist with maintenance in troubleshooting and with designers in making additions to the original building.

Reading Drawings

In the drawings, which are sometimes called prints or plans, there are various elevations shown, such as the front, side, and sectional. A plan is a horizontal view of an object. An elevation is a vertical view of an object.

Projected Views

The various views are projected on imaginary projection planes, similar to the projection of a picture on glass. To illustrate the first or front view, place a clear pane of glass in front of the object with the glass parallel to the surface of the object being projected. Fig. 9-1 shows a simple building with a shed roof. In front of the building is the pane of glass marked *V*, representing a vertical plane.

When an observer looks through the glass directly at the front of the object from a considerable distance, he will see only one side,

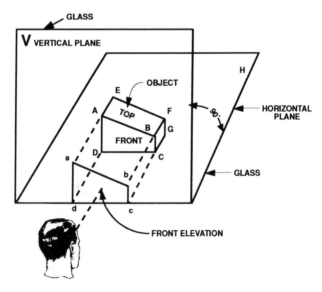

Fig 9-1. Rectangular object resting on a horizontal plane and facing a vertical plane.

in this case the side marked *ABCD*. The rays of light falling upon the object are reflected into the eyes of the observer, and in this manner he *sees* the object. The pane of glass (vertical plane) is placed so that the rays of light from the object will pass through the glass in *straight parallel lines* to the eyes of the observer.

The rays of light from points *ABCD* of the building pass through the glass at points *a, b, c,* and *d*. If these points *a, b, c,* and *d* are connected by lines, a view of the object as seen from the front is obtained, which is called *front elevation.*

The front elevation is identical in shape and size with the front side *ABCD* of the object; that is *ab* = *AB*; *bc* = *BC*, etc.; angle *dab* = angle *DAB*; angle *abc* = angle *ABC*, etc.

Top View or Plan

For this view, place a pane of glass in a horizontal position above the building which is resting on the horizontal plane (Fig. 9–2). Now, looking at the object directly from above, the rays of light from corners *AEFB* of the top pass through the glass at points *aefb*. If these points *aefb* are connected by lines, a view of the ob-

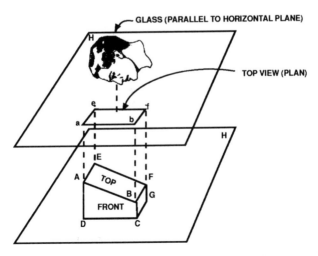

Fig. 9-2. Projection of the top of an object to obtain top view.

ject as seen from the top is obtained, which is called the *top view*, or preferably *plan*.

Right-End View (Elevation)

A pane of glass is placed to the right of the building in a vertical position and parallel to the right side *BFGC* of the building (Fig. 9–3). Here the pane of glass is marked *P*, which means *profile plane*, or plane from a side projection. Looking at the building directly from the right side (as in position *S*), the rays of light from corners *BFGC* of the upper left-hand side (from points *AE*) pass through the glass at points *bfgc* and *ae*. If these points are connected by lines, a view of the object as seen from the right side is obtained, which is called the *right-side view*, or preferably *right-end elevation*.

The shape of the object is such that the entire visible surface does not lie in a plane parallel to the projection plane. The points *A* and *E*, though located at the other end of the object, are visible and accordingly form part of the right-end view. Figure *aefb* does not show the top in its true size because it is projected obliquely instead

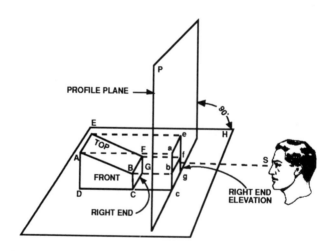

Fig. 9-3. Projection of the right end of an object illustrating right profile plane.

of at 90°. An oblique projection makes an object appear smaller than its real size.

Left-End View (Elevation)

With a pane of glass shifted to the left side of the object (Fig. 9–4), and the building viewed directly from the *left* side (as position S), the rays of light from corners *ADHE* of the left side pass through the glass at points *adhe*. If lines connecting these points are drawn on the glass, a left side view of the object is obtained. But the edge *FB* at the other end is invisible. It is shown by a dotted line connecting *f* and *b* projected from *F* and *B*. The completed drawing is then called a *left-side view*, or preferably a *left-end elevation*.

Sections

Most buildings are so complex they cannot be clearly represented by a plan and elevation alone. In such a case, the parts which do not appear properly in these drawings are better represented by a *section*, or *sectional views*. A cross section is a drawing of a building showing that part cut by a plane (Fig. 9–5).

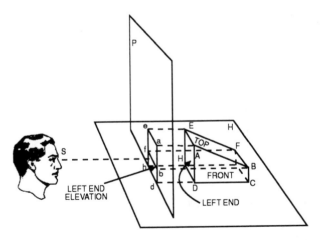

Fig. 9–4. Projection of the left end of an object illustrating left profile plane.

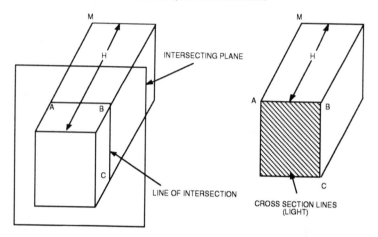

Fig. 9–5. Pictorial view of a building illustrating cross section.

Directions of View for Sectional Views—For an unsymmetrical object, it is important to know the direction in which the sectional view is viewed. This is indicated by arrows at the end of the line representing the cutting plane (Fig. 9–6). The arrows *AA* indicate that the object is viewed in the direction of point *L* (toward the smaller end of the object), and the arrows *BB* (in the direction of *R*) toward the larger end.

The Scale

Scale is the ratio between the actual size of the object and the size that it will be drawn. The scale is usually expressed on a drawing as full size, half size, quarter size, etc., or it might be expressed as 1 inch = 1 foot; 1 inch = 100 feet; 1 inch = 1000 feet, or any other proportion that might be necessary to use. The scale is printed on the drawing.

On a full-size drawing, the object and drawing are of the same size. When the drawing is marked half size, the object is twice the size of the drawing. Thus the drawing of an object is shown full size, half size, and quarter size (Fig. 9–7). If the building's height is represented by *H* and its diameter by *D*, then these dimensions will

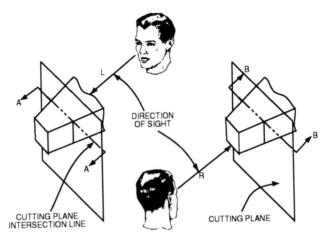

Fig. 9-6. Direction in which a sectional view is viewed is indicated by the direction of arrows AA, at the ends of the cutting plane intersection line.

be the same for the full-size drawing. That is, $H = h$; $D = d$. For the half-size drawing, $h' = \frac{1}{2} H$; $d = \frac{1}{2} D$. Similarly, for the quarter-size drawing, $h'' = \frac{1}{4} H$; $d'' = \frac{1}{4} D$.

From this it is seen that when the length of any edge on the drawing is made the same as the length of the corresponding edge on the object, the drawing is marked *full size* (sometimes *actual size*). If the length of any line on the drawing is half the length of the corresponding line on the object, the drawing is *half size*.

The scale is smaller than the building. In using prints, the size is important since they are used in the field, and if too large, they may not be easily handled.

In the case of a building, it would be impossible to have a print as large as the building; thus it is necessary to cut the print down in size. This necessitates the use of a scale. In the drawing of a building the building designers usually express scale as 1 inch = 1 foot; $\frac{1}{2}$ inch = 1 foot; etc. This would indicate that one inch on the drawing would be equal to 1 foot on the actual structure, or $\frac{1}{2}$ inch on the drawing would equal 1 foot on the actual structure respectively. The architect uses the architects' scale. This is laid out in inches, $\frac{1}{8}$'s, $\frac{1}{4}$'s, $\frac{1}{2}$'s, etc.

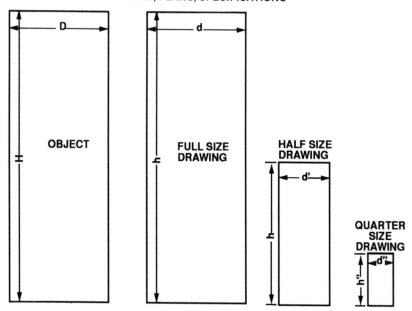

Fig. 9-7. A drawing of an object to different scales; full size, half size, and quarter size.

To lay off a distance of 2' 6", place the ¾ scale with division 2 at the given point A, then the zero division on the scale will be at a distance of 2 feet (Fig. 9–8). Since the end space is divided into twelfths, each division represents one inch on the ¾ scale. Therefore, measuring off six divisions indicates that $AB = 2'$ 6". Notice the difference in actual length of this measurement on the 1" = 1' scale (Fig. 9–9).

The scale on the original would not apply to a reproduction made by photocopying. The scale on the original (Fig. 9–10) should be crossed out and a *graphic scale of proportions* corresponding to the reproduction added.

Drawing Development

A knowledge of how architectural drawings are developed will help the carpenter and builder read drawings with more complete

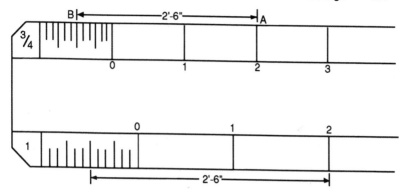

Fig. 9-8. A typical architect's scale; detail showing ¾ and 1 inch to the foot.

understanding. Architectural drawings are a means of transferring the thoughts of the designer to the builders and craftsmen whose responsibility it is to construct the building. Graphic symbols are used to locate specific features and where they are to be placed.

The different parts or drawings that are necessary to show the structure, such as the mechanical and electrical installations, are all shown graphically. Therefore, one must become familiar with these symbols, not only those of one particular trade, but those of all the trades. This is necessary so that complete coordination may

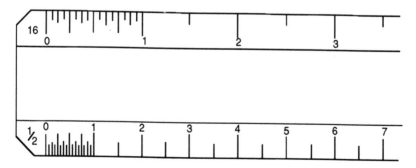

Fig. 9-9. A typical architect's scale detail with inch divisions instead of inches to foot.

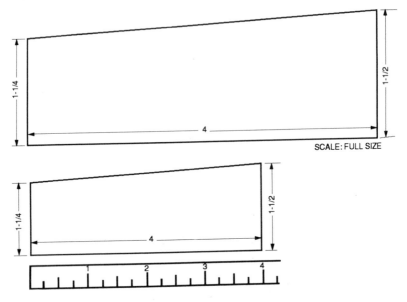

SCALE: FULL SIZE

Fig. 9-10. Graphic reproduction scale.

be reached between the various trades. In the construction of a building, time and money may be saved by representatives of the different trades sitting down together and going over the plans and laying out the pattern to be followed. No trade can work independently of the other. If this is attempted, confusion is created and some work will have to be done over to make all parts of the scheme fit together. During construction, the general contractor, the plumber, the steel workers, and the electrical and mechanical contractors must lay out the work together and determine from the drawings who installs what, where, and when.

On the job, the architect may have a representative present to assist in coordinating the work and in making the decisions that may be required. In the designing of the building, the owner or builder will often draw a rough sketch after which he/she sits down with the architect to discuss what the owner will need and want in a building design. The requirements are noted as to space, machinery, electrical loads, numbers of persons that will occupy the building, and what the future requirements might be. The owner will sometimes have a rough sketch of his ideas (Fig. 9–11, 9–12). These need not be drawn to scale or with any degree of accuracy; they are merely his ideas of what he might want. In this discussion, no elaborate plans are given. Simple plans are used, as they show the intent, and fall into line with more detailed plans.

Notice that there are no details shown, merely a sketch of the

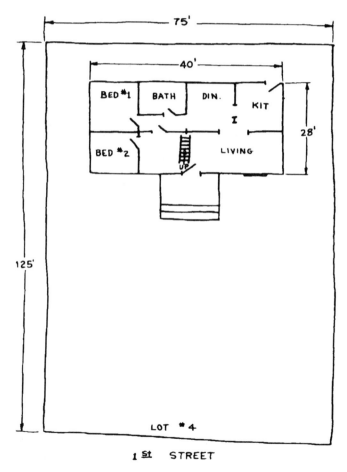

Fig. 9–11. Owner's sketch of the first floor of a residence.

spaces to be enclosed (Fig. 9–11, 9–12). After the sketch is drawn, the architect and owner can sit down and discuss details, at which time no doubt another free-hand sketch will be drawn with more details. When a tentative solution is reached, the architect will make a preliminary drawing (Fig. 9–13).

When the owner has signed construction papers, the architect will start drawing up the final plans and all details. There are many preliminary things to do, such as surveying the land to see how much excavating will be required. The location of the prop-

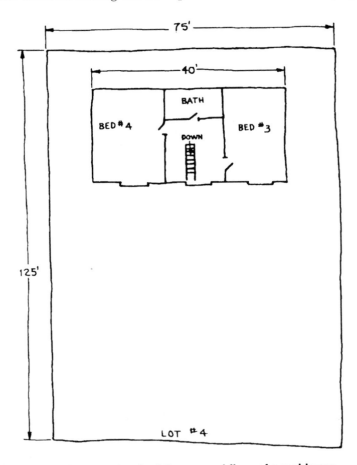

Fig. 9-12. Owners sketch of the second floor of a residence.

FIRST FLOOR PLAN

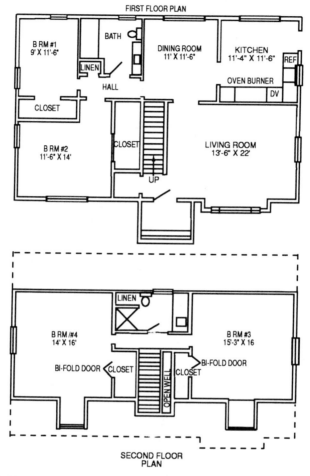

Fig. 9-13. Architect's drawing (preliminary), without dimensions, of the owner's idea.

erty lines and the general drainage plan for the immediate vicinity must be considered, and the water, sewer, gas, telephone and power lines that exist must also be considered. Local regulations regarding types of construction permitted, setbacks, etc., all must be taken into consideration.

The final drawings are sent to the plan checkers of the Inspec-

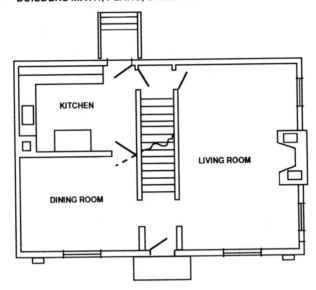

Fig. 9-14. A typical first-floor plan.

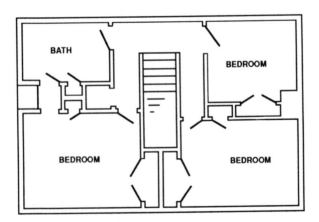

Fig. 9-15. A typical second-floor plan.

tion Departments having jurisdiction. Here, they are checked to see that they conform to local codes. Corrections are noted or the plans are approved. Most specifications that accompany plans put the burden of following local codes that are applicable on the tradesmen and contractors. When questions arise, such as an electrical or mechanical problem, the contractor involved takes these problems up with the architect or his assistant who, in turn, takes them to the engineer that has performed the design work.

There are various methods of bidding on plans. Sometimes, the general contractor gives the entire bid, but calls for bids from subcontractors. This method has some advantages in that the general contractor is responsible for the entire job.

At times, the general contractor and each subcontractor bid

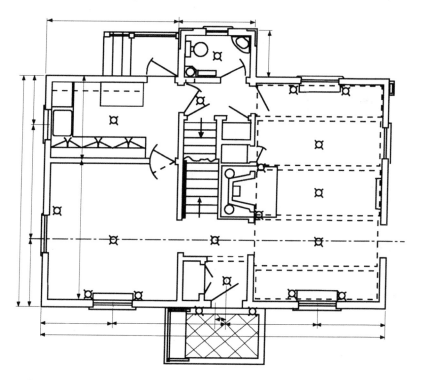

Fig. 9-16.   A typical first-floor plan with dimension lines.

their parts separately. This method has an advantage; the general contractor's percentage is removed from the subcontractors' bidding, and the owner has more control over who gets the bids. The details of the architect's sketches for the preliminary part are not important to the mechanic; therefore, they will not be covered in this book. What the mechanic is interested in is how he has to perform his part of the work. In this chapter, most of our efforts have been on details covering construction, and most of the drawings shown have followed this type of pattern.

Fig. 9–19 illustrates a typical detail of the basement wall, footings, and floor. Fig. 9–20 shows a typical detail of the floor joists, brick veneering, etc., as it attaches to the foundation shown in Fig. 9–19. A typical floor plan of the first floor and second floor is shown in Figs. 9–14 and 9–15. Figs. 9–16 and 9–17 show a typical floor plan of a first and second floor, with dimensions added. Fig. 9–18 is a typical plan of a basement, with dimensions added.

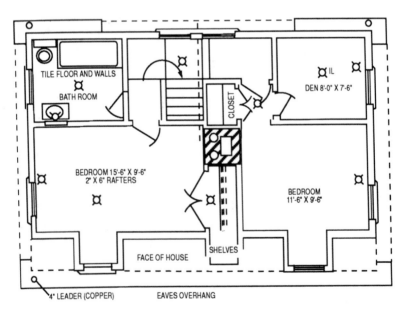

Fig. 9-17.   A typical second-floor plan with dimensions.

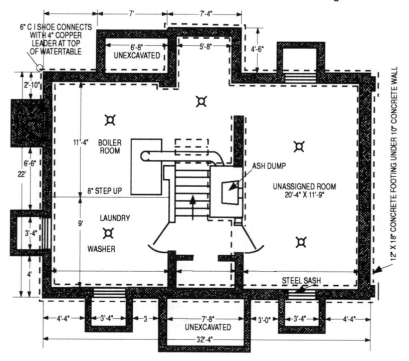

Fig. 9-18. A typical basement plan with dimensions.

Graphic Symbols

In architectural drawings, a form of shorthand is used to illustrate what is to be installed and at what point or location the installation will be in the building. These are commonly known as *graphic symbols*.

Symbols are used in drawings to represent various parts and systems; you must also become familiar with these. Each trade has its own symbols, and the craftsmen of each trade should learn to recognize the symbols of all the other trades. For example, the electrician should understand the plumber's symbols, the plumber should understand the carpenter's symbols, etc. In this way, each

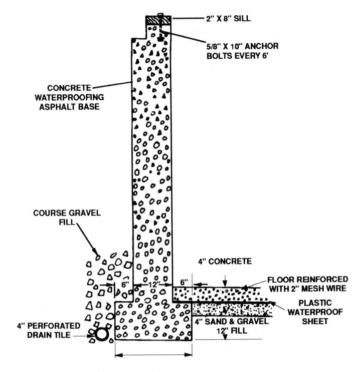

Fig. 9-19. Detail of basement walls, footings, and floor.

craftsman will know what obstacles may be encountered in the work and will be better prepared to cope with them.

The symbols shown are standards for the construction industry. However, you will find that some designers or individual institutions will deviate from these standards. Where this is done, a legend showing what the symbols mean should be added to the drawings.

A drawing consists of many different kinds of lines, each having its own purpose. Certain characteristic lines are used to convey different ideas, and the drafting practice has been rather well standardized as to the use of lines to avoid confusion in reading drawings. A good working drawing is as simple as possible, using only such lines as are necessary to give all of the required information.

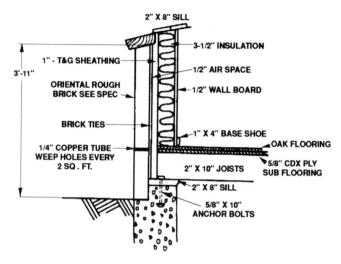

Fig. 9-20.   Details of floor, brick veneering, etc.

Moreover, the reader will not have to puzzle over a mass of lines which will complicate the drawing. The same thing holds true for dimensions and other data. A good drawing is accurate and complete, though simple, and is therefore easily understood ("read") by the carpenter.

The lines generally used on drawings are shown in Fig. 9-21.

Walls of frame buildings are represented on floor plans by two parallel lines spaced at a distance apart equal to the wall thickness, *A* in Fig. 9-22. Masonry walls are shown on a floor plan by cross sectional lines, as shown in *B*. Walls of all types of construction may

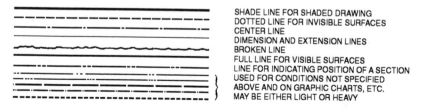

SHADE LINE FOR SHADED DRAWING
DOTTED LINE FOR INVISIBLE SURFACES
CENTER LINE
DIMENSION AND EXTENSION LINES
BROKEN LINE
FULL LINE FOR VISIBLE SURFACES
LINE FOR INDICATING POSITION OF A SECTION
USED FOR CONDITIONS NOT SPECIFIED
ABOVE AND ON GRAPHIC CHARTS, ETC.
MAY BE EITHER LIGHT OR HEAVY

Fig. 9-21.   Various lines used in drawing.

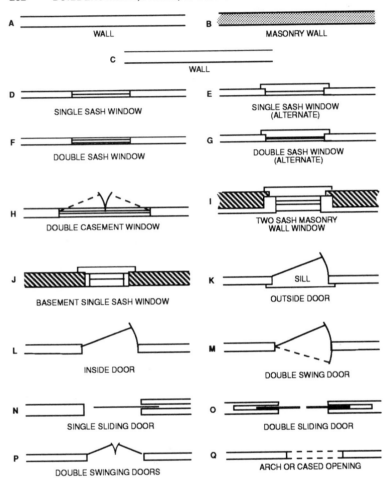

Fig. 9–22. Symbols for walls, windows, and openings.

also be shown as in *C*, by heavy dark lines which save time in drawing and give a better print.

There are many and varied types of window construction, the symbols for some being shown in *D* through *J*. The specifications should show the materials and types of construction, thickness of glass, and type of glass to be used. These details may be listed as a supplement to the specifications, or on the drawings if there is room.

There are many types of doors which will be used, the symbols for some of these being shown in K through Q. There may also be special doors used, and where this is called for, a drawing showing the detail should accompany the main drawing. Detailed sketches or drawing inserts should show all details of sills, especially where masonry construction is to be used. In drawings, the dashed line should be avoided where it is intended to indicate some part that is in view; the dashed line is ordinarily intended to represent some hidden feature or part.

A few conventions or symbols for chimneys and fireplaces are shown in Fig. 9–23. There may be special features which should be shown in additional drawings. In each case where details are required, a notation should be added referring to the detail drawings.

Stairs must be identified as to their direction, and whether they are boxed or open. Some methods of identification are shown in Fig. 9–24. Arrows show the direction of the stairs.

Symbols for the identification of materials are shown in Fig. 9–25.

If confused as to an abbreviation in reading a drawing, con-

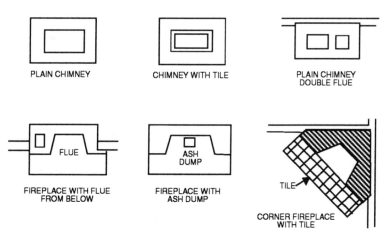

Fig. 9–23. Symbols for chimneys and fireplaces.

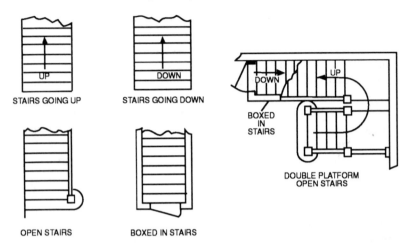

Fig. 9-24. Symbols for stairs.

sider the nature of the work, which will be helpful in interpreting the abbreviations. It should be understood that these abbreviations relate only to one part of the subject. For every field, such as carpentry, electrical work, etc., there are many conventions relating to each individual field. The following are common abbreviations used in construction drawings.

Common Abbreviations on Prints

Access Door	AD	Asbestos Board	AB
Access Panel	AP	Asphalt	ASPH
Acoustic	ACST	Asphalt Tile	A Tile
Aggregate	AGGR	Automatic Washing Machine	
Aluminum	AL		AWM
Anchor Bolt	AB	Basement	BSMT
Angle	ANG	Bathroom	B
Apartment	APT	Bath Tub	BT
Area	A	Beam	BM
Area Drain	AD	Bearing Plat	BRG PL
Asbestos	ASB	Bedroom	BR

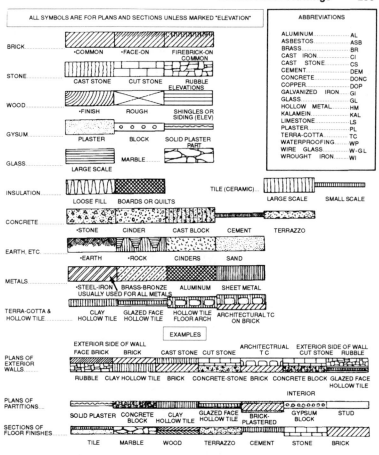

Fig. 9-25. Symbols for various materials.

Blocking .... BLKG	Boundary .... BDY
Blueprint .... BP	Brass .... BR
Boiler .... BLR	Broom Closet .... BC
Bolts .... BT	Building Line .... BL
Book Shelves .... BK SH	Cabinet .... CAB

Caulking	CLKG	Garage	GAR
Casing	CSG	Gas	G
Catch Basin	CB	Gauge	GA
Cellar	CEL	Gypsum	GYP
Cement Floor	CEM FL	Hall	H
Center	CTR	Hardware	HWD
Center to Center	C to C	Hose Bibb	HB
Center Line	C/L	Hot Air	HA
Ceramic	CER	Hot Water Tank	HWT
Channel	CHAN	I Beam	I
Cleanout	CO	Inside Diameter	ID
Clear Glass	CL GL	Insulation	INS
Closet	CLO	Iron	I
Cold Air	CA	Kitchen	K
Cold Water	CW	Knocked Down	KD
Conduit	CND	Landing	LDG
Counter	CTR	Lath	LTH
Cubic Feet	CU FT or FT^3	Living Room	LR
Detail	DET	Main	MN
Diagram	DIAG	Matched and Dressed	M & D
Dining Alcove	DA	Maximum	MAX
Dining Room	DR	Medicine Cabinet	MC
Double Acting Door	DAD	Minimum	MIN
Double Strength Glass	DSG	Miscellaneous	MISC
Drain	D or DR	Mixture	MIX
Electric Panel	EP	Mortar	MOR
End To End	E to E	On Center	OC
Excavate	EXC	Pantry	PAN
Expansion Joint	EXP JT	Partition	PARTN
Finished Floor	FIN FL	Plaster	PLAS
Firebrick	FRBK	Plate	PL
Fireplace	FP	Porch	P
Fireproof	FPRF	Precast	PRCST
Flooring	FLG	Prefabricated	PREFAB
Flush	FL	Pull Switch	PS
Footing	FTG	Radiator	RAD
Foundation	FND	Recessed	REC
Frame	FR	Refrigerator	REF

Register .... REG	Stairs .... ST
Revision.................... REV	Standard..................... STD
Riser .... R	Switch .... SW or S
Rivet........................... RIV	Storage........................ STG
Room .... R or RM	Telephone.................... TEL
Rubber Tile .... R Tile	Thermostat .... T or THERMO
Screen .... SCR	Tongue and Groove .... T&G
Section .... SECT	Unexcavated........... UNEXC
Sewer........................ SEW	Vent V
Shelving................ SHELV	Vinyl Tile .... V Tile
Shower........................ SH	Washroom .... WR
Single Strength Glass ...... SSG	Water .... W
Sink....................... S or SK	Water Closet................. WC
Soil Pipe .... SP	Water Heater .... WH
Square Feet ...... SQ FT or FT^2	Weatherstripping........... WS

Summary

Drawings are a means of communication between the designers and the builders. They are used on the construction site for reference by the tradespeople and as a record of how the building was actually built. The building is represented with various views such as plans and elevations. Complex views of the building are shown with sections. Sections are like a slice of the building that opens up the interior to a view. A floor plan is a horizontal slice.

Drawings are made to scale. There is a consistent relationship between the length of lines on the drawing and the size of the building and its parts.

Drawings are developed in phases that allow the designer to set the owner's requirements down on paper and show how the design can be changed to meet various functional and legal requirements.

A drawing is made up of symbols that represent the various parts of the building and where those parts will be installed during construction. The symbols are usually standardized so everyone knows what they mean. Simple symbols like lines are used to define the basic shape of the structure. More complex symbols indicate doors, windows, stairways and other common parts of the building.

Review Questions

1. What is the basic purpose of using drawings in construction of a building? How is this accomplished?
2. Why are the drawings updated during construction?
3. What part of a building does an elevation show?
4. What part of a building does a section show?
5. Why are drawings reduced in size by scaling?
6. Who, besides the carpenters and builders, uses the drawings of a construction project?
7. What do graphic symbols do in a drawing?
8. Draw the symbol for a door and a window.

CHAPTER 10

Building Styles

The Two-Story New England Colonial House

Many houses of the Colonial style of architecture, built in the late 1600's, are still standing in the New England states.

Many of the original houses were much larger than the one shown built to accommodate the larger families, but certain architectural details were common in all of them (Fig. 10–1). One feature found in most of the original houses was moderate to steep roof slopes, often the one-third pitch shown in the illustration. This was necessary to allow the use of wood shingles or shakes, about the only roof covering then generally available. Narrow eaves, with little or no projection at the gables, had the lap siding cut against wide corner boards. Plank frames were used for the windows, and there were no casings or very narrow casings outside. Entrances, however, were usually elaborate, sometimes with finely scrolled and carved pediments. Good, authentic replicas of many of these entrances are obtained today. In the later and more pretentious houses built in this era, side lights were often used at the entrances, and sometimes the doors were double. The relatively wide pilasters

at the sides were usually fluted or molded. A type of door that originated in England was paneled, three high, two wide, with a small,
nearly square, pair of panels at the top. The rails in the upper part
of the door form the Christian cross.

The front of the house was symmetrical about the central entrance. Although there were some exceptions to this design, sym-
metry was the rule. Invariably, second-story windows were placed
directly over lower story openings. Small covered entrances were
uncommon, porches and verandas virtually unknown. The central
entrance hall was universal, with the stairway to the upper floor.
In the earlier and smaller Colonial houses the stairway was often
steep and tortuous, often with tricky winders, but in the later and
better homes the stairs were often beautiful, tastefully designed
and elaborately carved, with superb craftsmanship. Needless to
say, such stairs cannot be duplicated today.

While the basic design was rectangular, many of the Colonial
homes had attached ells or sheds. With a roof continuous down
over a one-story shed at the rear, this house becomes the well-

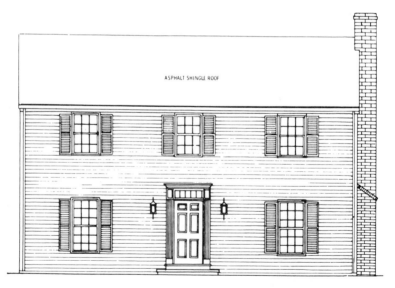

Fig. 10-1. Front elevation of a Colonial-type home.

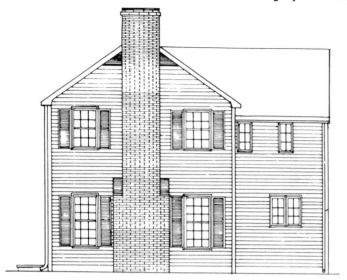

Fig. 10-2.   Right side elevation of a Colonial-type home.

known *salt-box*, with a claim to fame all its own. All types of Colonial homes use windows of small glass or lights, often rectangular, sometimes diamond-shaped. In the early days the muntins were often made of lead; later wood muntins were used. Large sheets of glass were almost unknown, and very expensive.

A predominant feature in the Colonial house is the *privacy* afforded, the privacy that is lacking, and so often deplored, in many modern designs. There are no unnecessarily large areas of glass to give one the eerie feeling of being spied upon at night.

Aside from the fact that this house is excellent in architecture, it need not be an expensive house (relative to other houses) to build. The downstairs bathroom is convenient, and the two complete bathrooms upstairs help prevent congestion (Figs. 10-3, 10-4). A complete basement is suggested for convenience to heating and plumbing, plus storage space. Included in the original plans is an adjacent garage with a covered breezeway connecting it with the outside door of the family room.

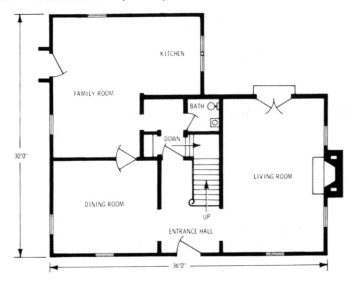

Fig. 10-3. First-floor plan of a Colonial-type home.

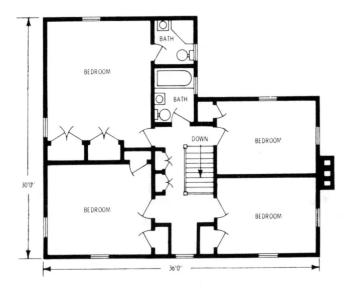

Fig. 10-4. Second floor plan of a Colonial-type home.

A House of Modern Architecture

Architects who have pride in their originality often prefer an unusual custom design (Fig. 10-5). This does not mean that a house of unusual appearance such as this one, no matter how good the appearance, can be placed on the odd lot in a street of Colonial houses and have a pleasing effect. Houses such as this one need a proper setting, preferably a rugged and individual setting.

The design is bold, and reflects the lively imagination of its designer. Usually modern practice makes use of large areas of glass, but these areas are effectively shaded by the wide overhangs, which make them much more acceptable than the large unprotected areas of glass in many modern designs.

The outstanding feature of this house is the roof. Although decidedly unusual, the pitches are regular, 45 degrees, and the roof is composed of simple intersecting planes. The roof framing is actually rather simple, for there are no *warped* planes. The architect has used diagonal lines on the roof instead of the commonly used vertical and horizontal accents. The effect is striking, especially on a sloping site, but working drawings are available which adapt it to level sites as well.

Although imaginative and unusual, the plan is in right-angled shapes for ease and economy of construction. The roof is formed of intersecting planes, with steep pitches. The roof treatment is important, as it dominates the entire view.

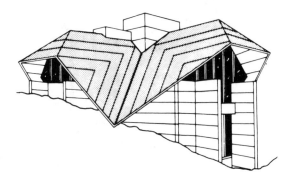

Fig. 10-5. A house of modern architecture.

In the plans, designations of rooms, and dimensions, are intentionally omitted (Figs. 10–6, 10–7). Naturally, the uses of the rooms will depend upon the contour of the ground. The large room with the fireplace with undoubtedly be used for a living room, depending upon whether the entrance is at the second-story level as it may be if the house is built on a sloping site, or at the first floor level, as it may be if the ground is level. The plan is flexible enough to allow considerable leeway in deciding upon the exact room arrangements, and their subsequent uses.

Specifications

In general, the walls are 12-inch lightweight concrete blocks, their cavities filled with insulation, with head joints cut smooth, and horizontal joints struck. The roof is sheathed with plywood (Fig. 10–8). With a roof so steep, many types of roof coverings may be used, but probably one of the modern types of mopped down

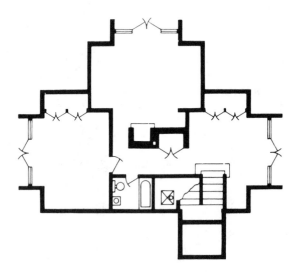

Fig. 10–6. The lower-floor level of the modern house. If the entrance is at this level, the room with the fireplace will certainly be used as the living room.

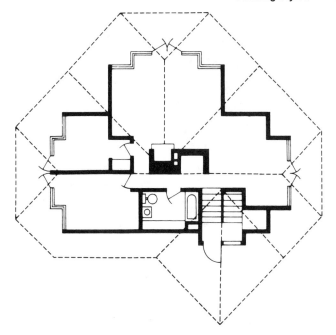

Fig. 10-7. The upper-floor level. If the entrance is at this level, the room with the fireplace will certainly be used as the living room.

roof covers would be most acceptable, with a granule-coated roll roofing as a cap sheet. The horizontal and diagonal roof lines are obtained by the use of battens.

The Contemporary House

This house is called contemporary architecture, for it embodies most of the features desired in homes of the present era (Fig. 10–9). Among these features we may enumerate the following:

- This house is a split-level, not a new idea, but always popular.
- It allows adequate windows in the lower-level rooms, and the stairway down is short.

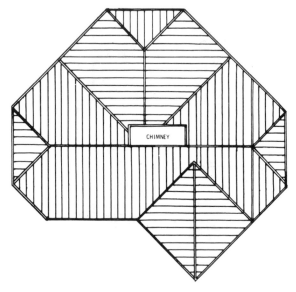

Fig. 10-8. The roof framing plan of the modern house.

Fig. 10-9. A split-level house.

- The living room is moderately large, and with the connected dining area it is ample for the needs of most moderately large families.
- The bedrooms are all of practical, usable size. The spare room in the basement is a useful stand-by.
- The access door to the two-car garage is convenient.
- The family room and the party room in the basement remove some of the inevitable activities of a large family from the living room.
- A concrete patio at the rear is convenient for the usual outdoor living activities of the modern family.

Construction Details

Glass has been used with discretion. Although the glazed areas are moderately large, they are not unreasonably so, at the expense of privacy and to the detriment of heating and air conditioning.

Architecturally, this house is susceptible to many variations of materials. In the house illustrated in Fig. 10–9, uncoursed native stone has been tastefully used, with vertical siding on the overhanging upper story.

The roof has sufficient pitch to fill its primary function which is to shed water, but it also gives a lower *spread out* look to the house.

It is possible with the proper use of a steel beam to enlarge the party or family room. This could be done by removing the wall between party room and spare room, to make one big room in an "L" shape (Fig. 10–10).

Utility Pole-Type Building

The pole-type building requires no foundation. It is mounted on round pole vertical members that are set in the ground and adequately anchored to resist uplift. For most buildings of appreciable size, the poles are set in round holes about 5 feet deep. Project-

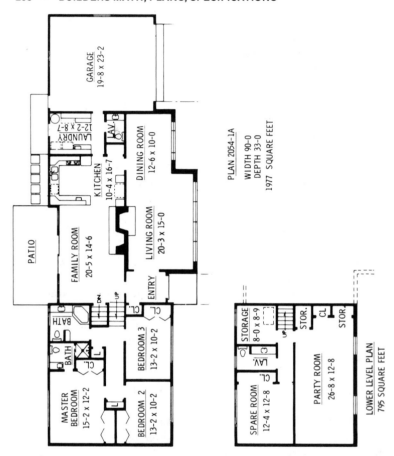

Fig. 10-10. Floor plans for the contemporary house.

ing nails or lag screws are placed in the pole 6 or 8 inches from the bottom, or holes are bored through which tight-fitting pieces of rod are driven to project a couple of inches on each side. After the poles are placed and positioned in the holes provided, about 12 inches of concrete is poured around the bottom ends and the holes filled with earth.

Satisfactory joining of rafters and other framing members to the round poles may be made to properly using any standard connecting device, including plain bolts, split-ring connectors, toothed-ring connectors and plain-shank common nails.

The idea of the pole-type building is not new. This type of construction was well known at least 75 years ago, but it was generally used only on inexpensive and temporary buildings. The modern pole building can hardly be placed in this category. The poles used are now pressure-treated with a preservative and have a long life even under most adverse conditions. Modern timber fastenings are far superior to anything that our forefathers had available.

Where rafters or trusses connect to the poles, it is preferable that the poles be flattened at the contacting surfaces. Cast-iron spike-grids with one side curved to conform to the round side of the pole have been satisfactorily used; however, simple bolts are also often used.

Summary

Many of the two-story New England houses built in the 1600's are still standing today. Features found in most original houses

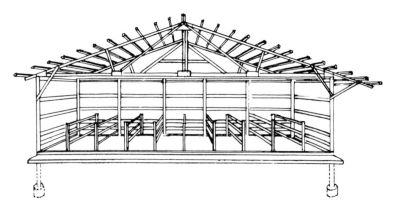

Fig. 10-11.   A pole-type building.

Review Questions

1. What type of house was constructed in the 1600's?
2. What is a split-level house?
3. What are some advantages in the split-level house?
4. Give a few advantages and disadvantages of a basement.
5. What is a pole-type building?

Reference

Wood Product and Lumber Associations

National Forest Products Association (NFPA)
1250 Connecticut Ave., N.W., Suite 200
Washington, DC   20036
202 463-2700
Publishes design specifications for construction with wood and booklets with technical data on the uses of wood.

California Redwood Association
405 Enfrente Dr., Suite 200
Novato, CA   94949
415 382-0662
Publishes pamphlets and data files of technical information on redwood.

National Hardwood Lumber Association (NHLA)
P.O. Box 34518
Memphis, TN   38184-0518
901 377-1818
Promotes research in hardwood timber utilization. Publishes hardwood grading rules.

STANDARD LUMBER ABBREVIATIONS

The following standard lumber abbreviations are commonly used in contracts and other documents for purchase and sale of lumber.

AD	air dried
ALS	American Lumber Standard
AST	antistain treated. At ship tackle (western softwoods)
AV or avg	average
AW & L	all widths and lengths
B1S	see EB1S, CB1S, and E&CB1S
B2S	see EB2S, CB2S, and E&CB2S
B&B, B&BTR	B and Better
B&S	beams and stringers
BD	board
BD FT	board feet
BDL	bundle
BEV	bevel or beveled
BH	boxed heart
BM	board measure
BSND	bright sapwood no defect
BTR	better
c	allowable stress in compression in pounds per square inch
CB	center beaded
CB1S	center bead on one side
CB2S	center bead on two sides
cft or cu. ft.	cubic foot or feet
CG2E	center groove on two edges
CLG	ceiling
CLR	clear
CM	center matched
Com	Common
CSG	casing
CV	center V
CV1S	center V on one side
CV2S	center V on two sides
DB Clg	double beaded ceiling (E&CB1S)
DB Part	double beaded partition (E&CB2S)
DET	double end trimmed
DF	Douglas-fir
DIM	dimension

DKG	decking
D/S, DS, D/Sdg	drop siding
D1S, D2S	See S1S and S2S
D&M	dressed and matched
D&CM	dressed and center matched
D&SM	dressed and standard matched
D2S&CM	dressed two sides and center matched
D2S&SM	dressed two sides and standard matched
E	edge
EB1S	edge bead one side
EB2S, SB2S	edge bead on two sides
EE	eased edges
EG	edge (vertical or rift) grain
EM	end matched
EV1S, SV1S	edge V one side
EV2S, SV2S	edge V two sides
E&CB1S	edge and center bead one side
E&CB2S, DB2S, BC&2S	edge and center bead two sides
E&CV1S, DV1S, V&CV1S	edge and center V one side
E&CV2S, DV2S, V&CV2S	edge and center V two sides
f	allowable stress in bending in pounds per square inch
FA	facial area
FAS	Firsts and Seconds
FBM, Ft. BM	feet board measure
FG	flat or slash grain
FJ	finger joint. End-jointed lumber using a finger joint configuration
FLG, Flg	flooring
FOHC	free of heart center
FOK	free of knots
FT, ft	foot or feet
FT. SM	feet surface measure
G	girth
GM	grade marked
G/R	grooved roofing

HB, H.B.	hollow back
HEM	hemlock
Hrt	heart
H&M	hit and miss
H or M	hit or miss
IN, in.	inch or inches
J&P	joists and planks
JTD	jointed
KD	kiln dried
LBR, Lbr	lumber
LGR	longer
LGTH	length
Lft, Lf	lineal foot or feet
LIN, Lin	lineal
LL	longleaf
LNG, Lng	lining
M	thousand
MBM, MBF, M. BM	thousand (feet) board measure
MC, M.C.	moisture content
MG	medium grain or mixed grain
MLDG, Mldg	molding
Mft	thousand feet
MSR	machine stress rated
N	nosed
NBM	net board measure
No.	number
N1E or N2E	nosed one or two edges
Ord	order
PAD	partially air dry
PART, Part	partition
PAT, Pat	pattern
Pcs.	pieces
PE	plain end
PET	precision end trimmed
P&T	posts and timbers
P1S, P2S	see S1S and S2S
RDM	random
REG, Reg	regular
Rfg.	roofing
RGH, Rgh	rough
R/L, RL	random lengths
R/W, RW	random widths

RES	resawn
SB1S	single bead one side
SDG, Sdg	siding
S-DRY	surfaced dry. Lumber 19 percent moisture content or less per American Lumber Standard for softwood
SE	square edge
SEL, Sel	select or select grade
SE&S	square edge and sound
SG	slash or flat grain
S-GRN	surfaced green. Lumber unseasoned, in excess of 19 percent moisture content per American Lumber Standard for softwood
SGSSND	Sapwood, gum spots and streaks, no defect
SIT. SPR	Sitka spruce
S/L, SL, S/Lap	shiplap
STD. M	standard matched
SM	surface measure
Specs	specifications
SQ	square
SQRS	squares
SR	stress rated
STD, Std	standard
Std. lgths.	standard lengths
SSND	sap stain no defect (stained)
STK	stock
STPG	stepping
STR, STRUCT	structural
SYP	southern yellow pine
S&E	side and edge (surfaced on)
S1E	surfaced one edge
S2E	surfaced two edges
S1S	surfaced one side
S2S	surfaced two sides
S4S	surfaced four sides
S1S&CM	surfaced one side and center matched
S2S&CM	surfaced two sides and center matched
S4S&CS	surfaced four sides and calking seam
S1S1E	surfaced one side, one edge
S1S2E	surfaced one side, two edges
S2S1E	surfaced two sides, one edge
S2S&SL	surfaced two sides and shiplapped
S2S&SM	surfaced two sides and standard matched
t	allowable stress in tension in pounds per square inch
TBR	timber

T&G	tongued and grooved
VG	vertical (edge) grain
V1S	see EV1S, CV1S, and E&CV1S
V2S	see EV2S, CV2S, and E&CV2S
WCH	west coast hemlock
WDR, wdr	wider
WHAD	worm holes a defect
WHND	worm holes no defect
WT	weight
WTH	width
WRD	western redcedar
YP	yellow pine

(Courtesy Forest Products Laboratory)

Characteristics of Wood Species

WOOD	CHARACTERISTICS	COMMENT
HARDWOODS:		
White or Gray Ash	Hard, heavy, springy; light reddish brown heart; sapwood nearly white.	Too hard to nail when dry.
Brown Ash	Brown heart, lighter sapwood.	Not a framing timber, but an attractive trim wood. Trees often wind shake so badly that the heart is entirely loose. Attractive veneers are sliced from stumps and forks.
Red Gum	Moderately heavy, interlocking grain; warps badly in seasoning; heart is reddish brown, sapwood nearly white.	Sapwood may be graded out and sold as white gum, heartwood as red gum, or together as unselected gum. Cuts into attractive veneers.
Hickory	Open-grained, tough, strong. Not rot-resistant.	Almost impossible to nail when dry.
Black Locust	Heavy, hard, strong; heartwood exceptionally durable.	Not a framing timber. Used mostly for posts and poles.
Hard Maple	Heavy, strong, hard and close-grained; color light brown to yellowish.	Used mostly for wear-resistant floors and for furniture. Circularly growing fibers cause the attractive "birds-eye" grain in some trees.
Soft Maple	Softer and lighter than hard maple; lighter colored.	Used for much the same purposes as hard maple, but not nearly so desirable.
White Oak	Hard, heavy, tough, strong, and somewhat rot-resistant. Brownish heart, lighter sapwood.	Used for trim and flooring; hardwood framing timbers.
Red Oak	———	Good framing timber, but not rot-resistant.

WOOD	CHARACTERISTICS	COMMENT
HARDWOODS, cont.:		
Black Walnut	Heavy, hard, strong; heartwood beautiful brown, sapwood heavy white. Somewhat rot-resistant.	Used for fine furniture, interior trim, gunstocks.
Butternut	Sapwood light to brown, heartwood light chestnut brown. Moderately light, rather weak, not rot-resistant.	Used for cabinet work and interior trim.
SOFTWOODS:		
Northern White Cedar	Light brown heart, sapwood thin and nearly white. Light, weak, soft, decay-resistant.	Holds paint well.
Western Red Cedar	Light, soft, straight-grained, small shrinkage. Heartwood is light brown, extremely rot-resistant. Sapwood quite narrow, nearly white.	Used for shingles, siding, boat building. Holds paint well.
Eastern Red Cedar, or Juniper	Pungent, aromatic odor. Red or brown heartwood, extremely rot-resistant, white sapwood.	Used for lining clothes closets and chests; also for fenceposts. Said to repel moths.
Cypress	Moderately light, close-grained; heartwood red to nearly yellow, sapwood nearly white.	Does not hold paint well, but otherwise good for siding and outside trim. Good looks also make it suitable for inside trim.
Eastern Hemlock	Heartwood pale brown to reddish, sapwood not distinguishable from heart. Brittle, moderately weak, not at all durable.	May be badly wind shaken. Used for cheap rough framing veneers.
Western Hemlock	Heartwood and sapwood almost white with purplish tinge. Moderately strong, not durable.	Mostly used for pulpwood.

WOOD	CHARACTERISTICS	COMMENT
SOFTWOODS, cont.		
Western White Pine	Creamy or light brown heartwood, sapwood thick and white. Moderately light, moderately strong, easy to work.	Used for millwork and siding. Holds paint well.
Red or Norway Pine	Moderately strong and stiff, moderately soft. Heartwood pale red to reddish brown.	Used for millwork, siding, framing, and ladder rails.
Long-Leaf Southern Yellow Pine	Heavy, hard, and strong; not durable in contact with soil. Sapwood takes preservative well.	One of the most useful timbers for light framing.
Short-Leaf Southern Yellow Pine	— — —	Used for light framing, sapwood for interior finish.
Douglas Fir	Strong, moderately heavy, splintery, splits easily	Used in all kinds of construction; rotary cut for plywood.
Yellow or Tulip Poplar	Red growth has yellow to brown heart. Sapwood and young trees are tough and white.	Not a framing lumber, but used for siding. Easy-working.
Redwood	Durable, rot-resistant, light, soft, moderately high strength. Heartwood reddish brown, sapwood white.	Paint "bleeds" through, used mostly for siding and outside trim.
Sitka Spruce	Light, soft, medium strong. Heartwood is light reddish brown, sapwood is nearly white, shading into the heartwood.	Used for boards, planingmill stock, boat lumber.
Eastern Spruce	Stiff, strong, hard, tough, moderately lightweight. Light color, little difference between heart and sapwood.	Used for pulpwood, framing lumber, millwork.

WOOD	CHARACTERISTICS	COMMENT
SOFTWOODS, cont.		
Engelmann Spruce	Straight-grained, light-weight, low strength. White with red tint. Extremely low rot resistance.	Used for dimension lumber, boards, pulpwood.
Tamarack or Larch	Yellowish-brown heart, sapwood white.	Used for boards, posts, and poles.

Index

AUDEL®

**Over a Century of Excellence
for the Professional
and
Vocational Trades and the Crafts**

Order now from your local bookstore
or use the convenient order form
at the back of this book.

AUDEL

These fully illustrated, up-to-date guides and manuals mean a better job done for mechanics, engineers, electricians, plumbers, carpenters, and all skilled workers.

CONTENTS

ELECTRICAL

House Wiring (Seventh Edition)
ROLAND E. PALMQUIST;
revised by PAUL ROSENBERG
*5 1/2 x 8 1/4 Hardcover 248 pp. 150 Illus.
ISBN: 0-02-594692-7 $22.95*
Rules and regulations of the current 1990 National Electrical Code for residential wiring fully explained and illustrated.

Practical Electricity
(Fifth Edition)
ROBERT G. MIDDLETON;
revised by L. DONALD MEYERS
*5 1/2 x 8 1/4 Hardcover 512 pp. 335 Illus.
ISBN: 0-02-584561-6 $19.95*
The fundamentals of electricity for electrical workers, apprentices, and others requiring concise information about electric principles and their practical applications.

Guide to the 1990 National Electrical Code
ROLAND E. PALMQUIST;
revised by PAUL ROSENBERG
*5 1/2 x 8 1/4 Hardcover 664 pp. 230 Illus.
ISBN: 0-02-594565-3 $24.95*
The most authoritative guide available to interpreting the National Electrical Code for electricians, contractors, electrical inspectors, and homeowners. Examples and illustrations.

New Book for 1991!
Installation Requirements of the 1990 National Electrical Code
PAUL ROSENBERG
*5 1/2 × 8 1/4 Hardcover 240 pp.
ISBN: 0-02-604941-4 $24.95*
Field guide for installation requirements makes understanding the 1990 Electrical Code simple while on the job. Applications and easy-to-understand tables make this the perfect working companion.

Mathematics for Electricians and Electronics Technicians
REX MILLER
*5 1/2 x 8 1/4 Hardcover 312 pp. 115 Illus.
ISBN: 0-8161-1700-4 $14.95*
Mathematical concepts, formulas, and problem-solving techniques utilized on-the-job by electricians and those in electronics and related fields.

Fractional-Horsepower Electric Motors
REX MILLER and
MARK RICHARD MILLER
*5 1/2 x 8 1/4 Hardcover 436 pp. 285 Illus.
ISBN: 0-672-23410-6 $15.95*
The installation, operation, maintenance, repair, and replacement of the small-to-moderate-size electric motors that power home appliances and industrial equipment.

Electric Motors (Fifth Edition)
EDWIN P. ANDERSON
and REX MILLER

5 1/2 x 8 1/4   Hardcover   696 pp.
Photos/line art
ISBN: 0-02-501920-1 $35.00

Complete reference guide for electricians, industrial maintenance personnel, and installers. Contains both theoretical and practical descriptions.

Home Appliance Servicing
(Fourth Edition)
EDWIN P. ANDERSON;
revised by REX MILLER

5 1/2 x 8 1/4   Hardcover   640 pp.   345 Illus.
ISBN: 0-672-23379-7 $22.50

The essentials of testing, maintaining, and repairing all types of home appliances.

Television Service Manual
(Fifth Edition)
ROBERT G. MIDDLETON;
revised by JOSEPH G. BARRILE

5 1/2 x 8 1/4   Hardcover   512 pp.   395 Illus.
ISBN: 0-672-23395-9 $16.95

A guide to all aspects of television transmission and reception, including the operating principles of black and white and color receivers. Step-by-step maintenance and repair procedures.

Electrical Course for Apprentices and Journeymen
(Third Edition)
ROLAND E. PALMQUIST

5 1/2 x 8 1/4   Hardcover   478 pp.   290 Illus.
ISBN: 0-02-594550-5 $19.95

This practical course in electricity for those in formal training programs or learning on their own provides a thorough understanding of operational theory and its applications on the job.

Questions and Answers for Electricians Examinations
(Tenth Edition)
Revised by PAUL ROSENBERG

5 1/2 x 8 1/4   Hardcover   316 pp.   110 Illus.
ISBN: 0-02-604955-4 $22.95

Based on the 1990 National Electrical Code, this book reviews the subjects included in the various electricians examinations—apprentice, journeyman, and master.

MACHINE SHOP AND MECHANICAL TRADES

Machinists Library
(Fourth Edition, 3 Vols.)
REX MILLER

5 1/2 x 8 1/4 Hardcover 1,352 pp. 1120 Illus.
ISBN: 0-672-23380-0 $52.95

An indispensable three-volume reference set for machinists, tool and die makers, machine operators, metal workers, and those with home workshops. The principles and methods of the entire field are covered in an up-to-date text, photographs, diagrams, and tables.

Volume I: Basic Machine Shop
REX MILLER

5 1/2 x 8 1/4   Hardcover   392 pp.   375 Illus.
ISBN: 0-672-23381-9 $17.95

Volume II: Machine Shop
REX MILLER

5 1/2 x 8 1/4   Hardcover   528 pp.   445 Illus.
ISBN: 0-672-23382-7 $19.95

Volume III: Toolmakers Handy Book
REX MILLER

5 1/2 x 8 1/4   Hardcover   432 pp.   300 Illus.
ISBN: 0-672-23383-5 $14.95

Mathematics for Mechanical Technicians and Technologists
JOHN D. BIES

5 1/2 x 8 1/4   Hardcover   342 pp.   190 Illus.
ISBN: 0-02-510620-1 $17.95

The mathematical concepts, formulas, and problem-solving techniques utilized on the job by engineers, technicians, and other workers in industrial and mechanical technology and related fields.

Millwrights and Mechanics Guide (Fourth Edition)
CARL A. NELSON

5 1/2 x 8 1/4 Hardcover 1,040 pp. 880 Illus.
ISBN: 0-02-588591-x $29.95

The most comprehensive and authoritative guide available for millwrights, mechanics, maintenance workers, riggers, shop workers, foremen, inspectors, and superintendents on plant installation, operation, and maintenance.

Welders Guide (Third Edition)

JAMES E. BRUMBAUGH

5 1/2 x 8 1/4   Hardcover   960 pp.   615 Illus.
ISBN: 0-672-23374-6 $23.95

The theory, operation, and maintenance of all welding machines. Covers gas welding equipment, supplies, and process; arc welding equipment, supplies, and process; TIG and MIG welding; and much more.

Welders/Fitters Guide

HARRY L. STEWART

8 1/2 x 11   Paperback   160 pp.   195 Illus.
ISBN: 0-672-23325-8 $7.95

Step-by-step instruction for those training to become welders/fitters who have some knowledge of welding and the ability to read blueprints.

Sheet Metal Work

JOHN D. BIES

5 1/2 x 8 1/4   Hardcover   456 pp.   215 Illus.
ISBN: 0-8161-1706-3 $19.95

An on-the-job guide for workers in the manufacturing and construction industries and for those with home workshops. All facets of sheet metal work detailed and illustrated by drawings, photographs, and tables.

Power Plant Engineers Guide
(Third Edition)

FRANK D. GRAHAM;
revised by CHARLIE BUFFINGTON

5 1/2 x 8 1/4   Hardcover   960 pp.   530 Illus.
ISBN: 0-672-23329-0 $27.50

This all-inclusive, one-volume guide is perfect for engineers, firemen, water tenders, oilers, operators of steam and diesel-power engines, and those applying for engineer's and firemen's licenses.

Mechanical Trades Pocket Manual (Third Edition)

CARL A. NELSON

4 x 6   Paperback   364 pp.   255 Illus.
ISBN: 0-02-588665-7 $14.95

A handbook for workers in the industrial and mechanical trades on methods, tools, equipment, and procedures. Pocket-sized for easy reference and fully illustrated.

PLUMBING

Plumbers and Pipe Fitters Library (Fourth Edition, 3 Vols.)

CHARLES N. McCONNELL

5 1/2 x 8 1/4   Hardcover   952 pp.   560 Illus.
ISBN: 0-02-582914-9 $68.45

This comprehensive three-volume set contains the most up-to-date information available for master plumbers, journeymen, apprentices, engineers, and those in the building trades. A detailed text and clear diagrams, photographs, and charts and tables treat all aspects of the plumbing, heating, and air conditioning trades.

Volume I: Materials, Tools, Roughing-In

CHARLES N. McCONNELL;
revised by TOM PHILBIN

5 1/2 x 8 1/4   Hardcover   304 pp.   240 Illus.
ISBN: 0-02-582911-4 $20.95

Volume II: Welding, Heating, Air Conditioning

CHARLES N. McCONNELL;
revised by TOM PHILBIN

5 1/2 x 8 1/4   Hardcover   384 pp.   220 Illus.
ISBN: 0-02-582912-2 $22.95

Volume III: Water Supply, Drainage, Calculations

CHARLES N. McCONNELL;
revised by TOM PHILBIN

5 1/2 x 8 1/4   Hardcover   264 pp.   100 Illus.
ISBN: 0-02-582913-0 $20.95

Home Plumbing Handbook
(Third Edition)

CHARLES N. McCONNELL

8 1/2 x 11   Paperback   200 pp.   100 Illus.
ISBN: 0-672-23413-0 $14.95

An up-to-date guide to home plumbing installation and repair.

The Plumbers Handbook
(Eighth Edition)

JOSEPH P. ALMOND, SR.;
revised by REX MILLER

4 × 6   Paperback   368 pp.   170 Illus.
ISBN: 0-02-501570-2 $19.95

Comprehensive and handy guide for plumbers and pipefitters—fits in the toolbox or pocket. For apprentices, journeymen, or experts.

Questions and Answers for Plumbers' Examinations
(Third Edition)
JULES ORAVETZ;
revised by REX MILLER

5 1/2 x 8 1/4 Paperback 288 pp. 145 Illus.
ISBN: 0-02-593510-0 $14.95

Complete guide to preparation for the plumbers' exams given by local licensing authorities. Includes requirements of the National Bureau of Standards.

HVAC

Air Conditioning: Home and Commercial (Fourth Edition)
EDWIN P. ANDERSON;
revised by REX MILLER

5 1/2 x 8 1/4 Hardcover 528 pp. 180 Illus.
ISBN: 0-02-584885-2 $29.95

A guide to the construction, installation, operation, maintenance, and repair of home, commercial, and industrial air conditioning systems.

Heating, Ventilating, and Air Conditioning Library
(Second Edition, 3 Vols.)
JAMES E. BRUMBAUGH

5 1/2 x 8 1/4 Hardcover 1,840 pp. 1,275 Illus.
ISBN: 0-672-23388-6 $53.85

An authoritative three-volume reference library for those who install, operate, maintain, and repair HVAC equipment commercially, industrially, or at home.

Volume I: Heating Fundamentals, Furnaces, Boilers, Boiler Conversions
JAMES E. BRUMBAUGH

5 1/2 x 8 1/4 Hardcover 656 pp. 405 Illus.
ISBN: 0-672-23389-4 $17.95

Volume II: Oil, Gas and Coal Burners, Controls, Ducts, Piping, Valves
JAMES E. BRUMBAUGH

5 1/2 x 8 1/4 Hardcover 592 pp. 455 Illus.
ISBN: 0-672-23390-8 $17.95

Volume III: Radiant Heating, Water Heaters, Ventilation, Air Conditioning, Heat Pumps, Air Cleaners
JAMES E. BRUMBAUGH

5 1/2 x 8 1/4 Hardcover 592 pp. 415 Illus.
ISBN: 0-672-23391-6 $17.95

Oil Burners (Fifth Edition)
EDWIN M. FIELD

5 1/2 x 8 1/4 Hardcover 360 pp. 170 Illus.
ISBN: 0-02-537745-0 $29.95

An up-to-date sourcebook on the construction, installation, operation, testing, servicing, and repair of all types of oil burners, both industrial and domestic.

Refrigeration: Home and Commercial (Fourth Edition)
EDWIN P. ANDERSON;
revised by REX MILLER

5 1/2 x 8 1/4 Hardcover 768 pp. 285 Illus.
ISBN: 0-02-584875-5 $34.95

A reference for technicians, plant engineers, and the homeowner on the installation, operation, servicing, and repair of everything from single refrigeration units to commercial and industrial systems.

PNEUMATICS AND HYDRAULICS

Hydraulics for Off-the-Road Equipment (Second Edition)
HARRY L. STEWART;
revised by TOM PHILBIN

5 1/2 x 8 1/4 Hardcover 256 pp. 175 Illus.
ISBN: 0-8161-1701-2 $13.95

This complete reference manual on heavy equipment covers hydraulic pumps, accumulators, and motors; force components; hydraulic control components; filters and filtration, lines and fittings, and fluids; hydrostatic transmissions; maintenance; and troubleshooting.

Pneumatics and Hydraulics
(Fourth Edition)
HARRY L. STEWART;
revised by TOM STEWART

5 1/2 x 8 1/4 Hardcover 512 pp. 315 Illus.
ISBN: 0-672-23412-2 $19.95

The principles and applications of fluid power. Covers pressure, work, and power; general features of machines; hydraulic and pneumatic symbols; pressure boosters; air compressors and accessories; and much more.

Pumps (Fifth Edition)

HARRY L. STEWART;
revised by REX MILLER

*5 1/2 x 8 1/4 Hardcover 552 pp. 360 Illus.
ISBN: 0-02-614725-4 $35.00*

The practical guide to operating principles of pumps, controls, and hydraulics. Covers installation and day-to-day service.

```
CARPENTRY AND
CONSTRUCTION
```

Carpenters and Builders Library
(Fifth Edition, 4 Vols.)

JOHN E. BALL;
revised by TOM PHILBIN

*5 1/2 x 8 1/4 Hardcover 1,224 pp. 1,010 Illus.
ISBN: 0-02-506450-9 $43.95*

This comprehensive four-volume library has set the professional standards for carpenters, joiners, a

Volume I: Tools

JOHN E. B
revise

*5 1/2 ver 384 pp. 345 Illus.
ISBN. 5-7 $10.95*

Volume Builders Math, Plans, Specifications

JOHN E. BALL;
revised by TOM PHILBIN

*5 1/2 x 8 1/4 Hardcover 304 pp. 205 Illus.
ISBN: 0-672-23366-5 $10.95*

Volume III: Layouts, Foundatio ming

JOHN E. BALL;
revised by TOM PH

*5 1/2 x 8 1/4 H 15 Illus.
ISBN: 0-672*

Volum Tools, Painting

JOH
revise HILBIN

*5 1/2 x 8 Hardcover 344 pp. 245 Illus.
ISBN: 0-672-23368-1 $10.95*

NEW EDITION FOR 1991–92

Complete Building Construction
(Second Edition)

JOHN PHELPS;
revised by TOM PHILBIN

*5 1/2 x 8 1/4 Hardcover 744 pp. 645 Illus.
ISBN: 0-672-23377-0 $22.50*

Constructing a frame or brick building from the footings to the ridge. Whether the building project is a tool shed, garage, or a complete home, this single fully illustrated volume provides all the necessary information.

Complete Roofing Handbook

JAMES E. BRUMBAUGH

*5 1/2 x 8 1/4 Hardcover 536 pp. 510 Illus.
ISBN: 0-02-517850-4 $29.95*

Covers types of roofs; roofing and reroofing; roof and attic insulation and ventilation; skylights and roof openings; dormer construction; roof flashing details; and much more.

Complete Siding Handbook

JAMES E. BRUMBAUGH

*5 1/2 x 8 1/4 Hardcover 512 pp. 450 Illus.
ISBN: 0-02-517880-6 $24.95*

This companion volume to the *Complete Roofing Handbook* includes comprehensive step-by-step instructions and accompanying line drawings on every aspect of siding a building.

Masons and Builders Library
(Second Edition, 2 Vols.)

LOUIS M. DEZETTEL;
revised by TOM PHILBIN

*5 1/2 x 8 1/4 Hardcover 688 pp. 500 Illus.
ISBN: 0-672-23401-7 $27.95*

This two-volume set provides practical instruction in bricklaying and masonry. Covers brick; mortar; tools; bonding; corners, openings, and arches; chimneys and fireplaces; structural clay tile and glass block; brick walls; and much more.

Volume 1: Concrete, Block, Tile, Terrazzo

LOUIS M. DEZETTEL;
revised by TOM PHILBIN

*5 1/2 x 8 1/4 Hardcover 304 pp. 190 Illus.
ISBN: 0-672-23402-5 $14.95*

Volume 2: Bricklaying, Plastering, Rock Masonry, Clay Tile

LOUIS M. DEZETTEL;
revised by TOM PHILBIN

*5 1/2 x 8 1/4 Hardcover 384 pp. 310 Illus.
ISBN: 0-672-23403-3 $14.95*

```
WOODWORKING
```

Wood Furniture: Finishing, Refinishing, Repairing
(Second Edition)

JAMES E. BRUMBAUGH

*5 1/2 x 8 1/4 Hardcover 352 pp. 185 Illus.
ISBN: 0-672-23409-2 $12.95*

A fully illustrated guide to repairing furniture and finishing and refinishing wood surfaces. Covers tools and supplies; types of wood; veneering; inlaying; repairing, restoring, and stripping; wood preparation; and much more.

Woodworking and Cabinetmaking

F. RICHARD BOLLER

5 1/2 x 8 1/4   Hardcover   360 pp.   455 Illus.
ISBN: 0-02-512800-0 $18.95

Essential information on all aspects of working with wood. Step-by-step procedures for woodworking projects are accompanied by detailed drawings and photographs.

```
MAINTENANCE AND
REPAIR
```

Building Maintenance
(Second Edition)

JULES ORAVETZ

5 1/2 x 8 1/4   Paperback   384 pp.   210 Illus.
ISBN: 0-672-23278-2 $11.95

Professional maintenance procedures used in office, educational, and commercial buildings. Covers painting and decorating; plumbing and pipe fitting; concrete and masonry; and much more.

Gardening, Landscaping and Grounds Maintenance
(Third Edition)

JULES ORAVETZ

5 1/2 x 8 1/4   Hardcover   424 pp.   340 Illus.
ISBN: 0-672-23417-3 $15.95

Maintaining lawns and gardens as well as industrial, municipal, and estate grounds.

Home Maintenance and Repair: Walls, Ceilings and Floors

GARY D. BRANSON

8 1/2 x 11   Paperback   80 pp.   80 Illus.
ISBN: 0-672-23281-2 $6.95

The do-it-yourselfer's guide to interior remodeling with professional results.

Painting and Decorating

REX MILLER and GLEN E. BAKER

5 1/2 x 8 1/4   Hardcover   464 pp.   325 Illus.
ISBN: 0-672-23405-x $18.95

A practical guide for painters, decorators, and homeowners to the most up-to-date materials and techniques in the field.

Tree Care (Second Edition)

JOHN M. HALLER

8 1/2 x 11   Paperback   224 pp.   305 Illus.
ISBN: 0-02-062870-6 $16.95

The standard in the field. A comprehensive guide for growers, nursery owners, foresters, landscapers, and homeowners to planting, nurturing and protecting trees.

Upholstering (Updated)

JAMES E. BRUMBAUGH

5 1/2 x 8 1/4   Hardcover   400 pp.   380 Illus.
ISBN: 0-672-23372-x $15.95

The esentials of upholstering fully explained and illustrated for the professional, the apprentice, and the hobbyist.

```
AUTOMOTIVE AND
ENGINES
```

Diesel Engine Manual
(Fourth Edition)

PERRY O. BLACK;
revised by WILLIAM E. SCAHILL

5 1/2 x 8 1/4   Hardcover   512 pp.   255 Illus.
ISBN: 0-672-23371-1 $15.95

The principles, design, operation, and maintenance of today's diesel engines. All aspects of typical two- and four-cycle engines are thoroughly explained and illustrated by photographs, line drawings, and charts and tables.

Gas Engine Manual
(Third Edition)

EDWIN P. ANDERSON;
revised by CHARLES G. FACKLAM

5 1/2 x 8 1/4   Hardcover   424 pp.   225 Illus.
ISBN: 0-8161-1707-1 $12.95

How to operate, maintain, and repair gas engines of all types and sizes. All engine parts and step-by-step procedures are illustrated by photographs, diagrams, and troubleshooting charts.

Small Gasoline Engines

REX MILLER and
MARK RICHARD MILLER

5 1/2 x 8 1/4   Hardcover   640 pp.   525 Illus.
ISBN: 0-672-23414-9 $16.95

Practical information for those who repair, maintain, and overhaul two- and four-cycle engines—including lawn mowers, edgers, grass sweepers, snowblowers, emergency electrical generators, outboard motors, and other equipment with engines of up to ten horsepower.

Truck Guide Library (3 Vols.)

JAMES E. BRUMBAUGH

5 1/2 x 8 1/4   2,144 pp.   1,715 Illus.
ISBN: 0-672-23392-4 $50.95

This three-volume set provides the most comprehensive, profusely illustrated collection of information available on truck operation and maintenance.

Volume 1: Engines

JAMES E. BRUMBAUGH

5 1/2 x 8 1/4   Hardcover   416 pp.   290 Illus.
ISBN: 0-672-23356-8 $16.95

Volume 2: Engine Auxiliary Systems

JAMES E. BRUMBAUGH

5 1/2 x 8 1/4   Hardcover   704 pp.   520 Illus.
ISBN: 0-672-23357-6 $16.95

Volume 3: Transmissions, Steering, and Brakes

JAMES E. BRUMBAUGH

5 1/2 x 8 1/4   Hardcover   1,024 pp.   905 Illus.
ISBN: 0-672-23406-8 $16.95

DRAFTING

Industrial Drafting

JOHN D. BIES

5 1/2 x 8 1/4   Hardcover   544 pp.   Illus.
ISBN: 0-02-510610-4 $24.95

Professional-level introductory guide for practicing drafters, engineers, managers, and technical workers in all industries who use or prepare working drawings.

Answers on Blueprint Reading
(Fourth Edition)

ROLAND PALMQUIST;
revised by THOMAS J. MORRISEY

5 1/2 x 8 1/4   Hardcover   320 pp.   275 Illus.
ISBN: 0-8161-1704-7 $12.95

Understanding blueprints of machines and tools, electrical systems, and architecture. Question and answer format.

HOBBIES

Complete Course in Stained Glass

PEPE MENDEZ

8 1/2 x 11   Paperback   80 pp.   50 Illus.
ISBN: 0-672-23287-1 $8.95

The tools, materials, and techniques of the art of working with stained glass.

Prices are subject to change without notice.